IN-SERVICE EDUCATION IN PRIMARY MATHEMATICS

Sister Mary Timothy Pinner, OSU
and
Hilary Shuard

Open University Press
Milton Keynes · Philadelphia

Open University Press
12 Cofferidge Close
Stony Stratford
Milton Keynes MK11 1BY. England
and
242 Cherry Street
Philadelphia, PA 19106, USA

First Published 1985

Copyright © Sister Mary Timothy Pinner
and Hilary Shuard 1985

All rights reserved. No part of this work may be reproduced in any form, by mimeograph or by any other means, without permission in writing from the publisher.

British Library Cataloguing in Publication Data

Pinner, Mary Timothy
  In-service education in primary mathematics.
  1. Mathematics—Study and teaching
  (Elementary)—Great Britain   2. Elementary
  school teachers—In-service training—
  Great Britain
  I. Title   II. Shuard, Hilary
  510'.7   QA135.5

ISBN 0-335-15023-3

Library of Congress Cataloging in Publication Data

Pinner, Mary Timothy, Sister.
  In-service education in primary mathematics.

  Bibliography: p.
  Includes index.
  1. Mathematics—Study and teaching (Primary)
2. Mathematics teachers—In-service training.
I. Shuard, Hilary.   II. Title.
QA135.5.P56 1985   372.7   84-18964

ISBN 0-335-15023-3 (pbk.)

Text design by Nicola Sheldon

Set by Mathematical Composition Setters Ltd.
Ivy Street, Salisbury, Wilts
Printed in Great Britain

Authors' acknowledgements

The project reported in this book was funded by a generous grant from British Petroleum, for the period 1979–82. Our warmest thanks are due to BP, and especially to Peter Coaker, who was unfailingly supportive and helpful throughout the lifetime of the project, and beyond.

Our thanks also go to Homerton College, which housed and encouraged the project, and especially to David Bridges, whose wide knowledge of INSET was constantly available to us. Penny Harrison acted as secretary to the project, and faithfully transcribed many hours of audio-tape of interviews.

Above all, our thanks go to the many people who took part in our work. Primary teachers, mathematics co-ordinators, Heads, Advisory Teachers, Teachers' Centre Wardens and Mathematics Advisers from a great number of LEAs, and INSET providers from a variety of institutions, gave most generously of their time and talked to us in great detail about their successes and their failures, their hopes and their fears, their understandings and their problems. They were united by a common enthusiasm to improve the teaching of mathematics in primary schools. We have not listed them; in a project based on case studies it is proper to take every precaution to preserve the anonymity of the informants. We hope we have succeeded; the case studies were read by two people who knew many of the INSET providers decribed in the case studies. We were reassured that they did not recognise their friends. If any clues have inadvertently slipped out, please accept our apologies.

Mary Timothy Pinner, OSU

Hilary Shuard

Contents

1	**The monitoring of INSET in primary mathematics**	1
	1.1 Introduction	1
	1.2 Previous work on in-service education	2
	1.3 The scope of the BP project	6
	1.4 Methodology	7
	1.5 The 'effectiveness' of in-service education	10
	1.6 A model of professional development	11
2	**Sources and types of INSET provision**	14
	2.1 Sources of provision	14
	2.2 Types of provision in Teachers' Centres	17
	2.3 Types of INSET provision in colleges	22
	2.4 Reasons given by colleges for their INSET preferences	24
	2.5 Discussion	27
3	**An experiment in monitoring — an LEA course for co-ordinators**	29
	3.1 Methodology	29
	3.2 Structure of the course	30
	3.3 The first day	30
	3.4 The second day	35
	3.5 The final day	36
	3.6 After the course	37
	3.7 Discussion	39
4	**Case studies — a group of two-day and three-day courses**	41
	4.1 Introduction to the case studies	41
	4.2 A weekend residential course for infant teachers	43

Contents

    4.3 A residential course for infant teachers   55
    4.4 A residential course for co-ordinators   63

5 **Case studies — short courses mounted by Teachers' Centres**   71
    5.1 Introduction   71
    5.2 A three-session course at a Teachers' Centre   71
    5.3 A classroom-based course   76
    5.4 A single session at a Teachers' Centre   83

6 **Case studies — longer courses**   86
    6.1 Introduction   86
    6.2 An intermittent course for mathematics co-ordinators   86
    6.3 A one-term full-time course in primary mathematics   99

7 **Case studies — other substantial commitments related to professional development**   112
    7.1 Introduction   112
    7.2 Open University courses and related INSET   113
    7.3 Informal INSET based on a project   124

8 **Interviews with teachers and INSET providers**   138
    8.1 Introduction   138
    8.2 Working with teachers in the classroom   139
    8.3 School-based INSET provided by a college   141
    8.4 The effects of courses   143
    8.5 Teachers' needs at different stages   145
    8.6 Needs of co-ordinators   146
    8.7 A teacher on professional development   148
    8.8 Discussion   150

9 **The individual teacher and INSET**   152
    9.1 The model of professional development   152
    9.2 Initiation   153
    9.3 Consolidation   154
    9.4 Integration   157
    9.5 Reflection   160
    9.6 The match between INSET courses and professional development   163
    9.7 Conclusion   166

10 **INSET in the service of the school**   169
    10.1 Available INSET provision   169
    10.2 The role of the mathematics co-ordinator   170
    10.3 The co-ordinator's professional development   171
    10.4 Feedback after courses   174
    10.5 Re-entry   176
    10.6 The influence of the Head   177

11	**Providers and INSET**	180
	11.1 Identification of needs	180
	11.2 Planning and advertisement	180
	11.3 Course timing and structure	182
	11.4 Discussion and practical activities	183
	11.5 Handouts	184
	11.6 Qualities of providers	185
12	**Conclusion: INSET as an agency of change**	187
	12.1 Introduction	187
	12.2 INSET as an agency of change	187
	12.3 The level of provision	189
	12.4 After Cockcroft, what next?	190
References		193
Index		195

CHAPTER 1

The monitoring of INSET in primary mathematics

1.1 Introduction

In 1979, British Petroleum funded a three-year Fellowship in Primary Mathematics to be held at Homerton College, Cambridge. For many years, BP has been a generous supporter of educational research and inquiry in Britain, and has funded projects that are intended to have a direct impact on classroom practice in areas that interest it, such as mathematics, the sciences and foreign languages. The present project was the first to be supported by BP in the field of primary education, and BP's interest sprang from the realisation that the foundations of children's mathematical development are laid in the primary school, but that many primary teachers claim no special expertise in mathematics, although they teach it daily to their classes. Development in the teaching of mathematics in primary schools might be expected to have an effect on children's overall understanding of mathematics throughout their schooldays. One of the major ways in which teachers can develop and reflect upon their teaching is through in-service education. Thus, the brief of the Fellowship was:

> to monitor the effectiveness of in-service education in mathematics for primary teachers.

The author of the proposal was Hilary Shuard, a member of the staff of Homerton College, Cambridge. She had been concerned in the initial training of primary teachers in mathematics for many years, and had taken part in a considerable amount of in-service work, both college-based and LEA-based. In-service work was not, in general, a major preoccupation of the colleges of education until after 1975, when the reorganisation of teacher education in England brought initial and in-service education for teachers closer together, and gave the colleges of education an increased role in in-service education. An important development for primary mathematics was the inception by the Mathematical Association in 1978 of a two-year part-time *Diploma in Mathematical Education* for teachers of the 5—13 age-range; courses for the Diploma were provided in many colleges, and this ensured the involvement of college tutors in in-service provision in primary mathematics at a fairly high academic level of study.

In 1978, also, the Cockcroft Committee was set up to inquire into the teaching of mathematics in primary and secondary schools; Hilary Shuard was a member of it, and chairman of its Primary Working Group. She became increasingly aware of the importance of in-service education in mathematics for primary teachers, and of our lack of knowledge about the most effective organisation of such training, and about the best use of scarce resources for the improvement of primary mathematics teaching.

These influences together contributed to the putting forward of a proposal that the existing provision of in-service education in primary mathematics should be monitored. The details of the proposal included:

> An objective of the monitoring will be to identify different styles of in-service work. Courses differ in length, target population, qualification aim ... other differences relate to the balance of content between mathematics and mathematical education. Some courses contain child studies and work on investigations. It is not known, however, what effects these elements have in practice on teachers' aims, attitudes and work in the classroom ... Courses also vary in their emphasis on different styles of teaching such as lecture, discussion, practical work, and the study of classroom materials and children's work ... An attempt will be made to identify and describe different styles of work and to assess their effects on the teachers who participate in them, and perhaps on their colleagues and pupils. The main method of working will be by observation and discussion with providers and participants in in-service education, and through interviews and visits to schools.

BP accepted the proposal, and Sister Mary Timothy Pinner, OSU, was appointed Fellow for the three-year period 1979—82, under the general direction of Hilary Shuard. Sister Timothy's background was in teacher education: she had previously been a member of the staff of Christ's College, Liverpool, and had been closely involved in working with primary schools in Liverpool to develop their mathematics teaching.

1.2 Previous work on in-service education

1.2.1 General studies of INSET

In-service education for teachers became a major topic of educational concern in the late 1970s. In 1978, an important national conference on in-service education was held at Bournemouth, and following this the Induction and In-service Subcommittee of the Advisory Council on the Supply and Training of Teachers (ACSTT) issued a paper, *Making INSET Work* (DES and Welsh Office 1978). This paper raised many issues involved in INSET under the four headings:

> Identify the main needs.
> Decide on and implement the general programme.
> Evaluate the effectiveness of this general programme.
> Follow up the ideas gained.

Although ACSTT intended this programme of action to be used by individual schools, it applies equally well to the whole INSET scene, and suggests that INSET continually needs to be evaluated and to adapt itself to changing needs, as the output from the fourth stage of the INSET cycle becomes the input to a new first stage. There have been only a few published studies of INSET in primary mathematics, but it is possible to apply principles drawn from more general discussions and evaluations of INSET to this specialised area.

Bolam (1979), in a paper given at the Bournemouth conference, provides an analytic framework for INSET evaluation, showing the interaction of the evaluator, the evaluation tasks and the target of INSET evaluation. In the present project, the evaluation target is that of the INSET *programme* at the level of the *providers*.

A useful book by Henderson (1978), *The Evaluation of In-service Teacher Training*, brought together existing expertise and experience of evaluation to point to lines for development; it both presented the historical context and explored questions of policy and methodology.

The OECD also commissioned a study of evaluation procedures in INSET in the United Kingdom; this project reported in 1980 (McCabe 1980), illustrating 'theories, principles and assumptions underlying current debates and practices' by means of case studies. McCabe noted that evaluation not only considered the value of INSET activities but also provided a new developmental facet to INSET itself:

> Teachers, whether presenting or attending courses, often see evaluation as both threat and burden ... Once resistance to evaluation is met it means that it ... is becoming worthwhile and a full part of the process of in-service education. And this is what it must become if it is to be effective. (McCabe 1980, p. 116)

Another collection of case studies is to be found in *In-service: the teacher and the school* (Donoughue 1981). In a preface to this collection, John Merritt points out that:

> It is our job to do what we can to help each individual child to learn how to cope with adult life in the best possible way that we can devise ... as there are no ready-made experts to teach teachers how to do this — nor can there be — all we can do is to find the most effective ways of sharing what relevant knowledge and expertise we already have.

> ... one of the most important issues raised by the case studies is that a teacher can no longer regard the teaching of a particular aspect of curriculum to a particular group of children as a purely personal responsibility.
>
> (Donoughue 1981, p. 9)

Donoughue presents examples of the INSET role of the school, the LEA, and regional and national agencies. Through reflection on attempts to evaluate INSET, it is hoped that it will be possible for teachers:

> to move away from a situation where INSET is seen as something organized *by* others *for* teachers, in which it is the 'others' who determine what is to be done. Rather, teachers, as professionals, should be taking the initiative.
>
> (Henderson, in Donoughue 1981, p. 251)

Two recent studies funded by the Schools Council have explored particular areas of the INSET scene; their titles are *Making the Most of the Short In-service Course* (Rudduck 1981) and *Teachers' Centres: a focus for in-service education?* (Weindling, Reid and Davis 1983). Rudduck found that:

> The present study reveals the deep sense of professional isolation that many teachers still feel; this makes attendance at outside in-service meetings a crucial aspect of professional life and an important condition of professional development ... The interaction between the short outside course and school-based in-service work needs to be carefully considered.
>
> (Rudduck 1981, p. 171)

Weindling *et al.* examined the role and function of Teachers' Centres; they found that among the main functions of a centre was the provision of short in-service courses; they also noted that:

> The data also suggested that different types of activity were required at different stages. Primary teachers, for example, tended to use centre courses in the early stages of their career and at a later stage became more involved with various types of working groups.
>
> (Weindling *et al.* 1983, p. 151)

1.2.2 INSET in mathematics

In the last section, general principles were considered; these may be applied in every curriculum area, including mathematics. However, there are also a few studies of INSET specifically in the field of mathematics, including two in primary mathematics.

In 1974, the National Council of Teachers of Mathematics in the United States began a project concerned with the problems of in-service education in mathematics. Teachers were surveyed and asked to identify in-service programmes they had found helpful, and providers were asked to describe their programmes so that critical variables in design might be identified. The outcome of this work was *An In-service Handbook for Mathematics Education* (Osborne 1977). In summary, it was found that some of the important ingredients contributing to the success of in-service programmes were:

1. the identification of needs and cooperative planning;
2. the encouragement of participation in in-service programs through adequate fundings and released time;
3. good leadership — the best is participatory but nevertheless leads to definite decisions and has the power to influence the system;
4. the team approach, which provides in-service programs for teachers who work together:
5. the proper combination of content and methods in the program;
6. having teachers take from in-service programs instructional materials that they can use in classrooms — the materials can be teacher-made or commercial;
7. the use of appraisal as an integral part of each in-service program;
8. regular and appropriate follow-up in the classroom and in subsequent progress review sessions.

(Rowan, Capehart and Sharpe, *in* Osborne 1977, p. 104)

The first substantial published British evaluation of INSET in primary mathematics was that of two-year part-time courses in colleges for the Diploma in Mathematical Education; this Diploma is validated by the Mathematical Association. The Diploma course aims 'to equip suitable and interested primary and middle school teachers to provide leadership in mathematics teaching within their schools.' The first students enrolled in 1978, and in 1980 the Department of Education and Science funded a two-year research project to monitor the progress of the Diploma and assess its effectiveness. The monitoring (Melrose 1982) was carried out by analysing data about the courses collected by means of visits and questionnaires, and from the Mathematical Association's records. A further in-depth study of forty teachers provided material for considering the effectiveness of the courses. The report comments on the students' background, the organisation of the courses at different centres, and on teaching and learning strategies; it also considers assessment and performance, and discusses the course outcomes for individuals and their schools. The final conclusion was that:

> The course, when it is well organised and taught and when conditions in the teachers' schools allow, can be effective in influencing for good the teaching of mathematics and the organisation of mathematics in schools.
>
> (Melrose 1982, p. 169)

The monitoring of INSET in primary mathematics

Edith Biggs, HMI, was a pioneer of the reform of primary mathematics in the 1960s and 1970s. She worked tirelessly to bring a new style of mathematics teaching to teachers, constantly touring the countryside, speaking and running practical workshops for teachers. However:

> It soon became apparent that the effects of the workshops on teachers were not lasting, with a few notable exceptions, mainly of those who had a special interest in mathematics and were prepared to sustain the impetus I had been able to provide.
>
> (Biggs 1983, pp. 5—6)

Consequently, in retirement Dr Biggs investigated the effects of another type of INSET in primary mathematics, on-site classroom support for teachers, using the methods of action research and case study. She found a number of reasons that many teachers are reluctant to commit themselves to change: these included a lack of personal experience of practical activities and investigations as a way of learning mathematics, and also a lack of understanding of the mathematics they were expected to teach. Among findings that she did not expect were the necessity for Heads to be actively involved in a project if lasting changes were to occur, and

> a persistent belief that it was more important for children to be able to perform calculations at an early stage than to understand what they were doing.
>
> (Biggs 1983, p. 193)

This book makes an important contribution to our understanding of the difficulty of changing mathematics teaching in the primary school.

1.3 The scope of the BP project

Henderson (1978) defines in-service training as 'structured activities designed, exclusively or primarily, to improve professional performance.' This definition is wide-ranging, and Henderson lists some activities that might be seen as part of it:

> It may involve attending a conference or listening to a lecture involving over 100 participants. It may involve full or part-time attendance at a specific course of instruction over a few hours, days, weeks, months or even years in the company of a few dozen colleagues. It may involve a joint problem-solving exercise, perhaps in the curricular or administrative field, with a small group of teachers from the same school or locality. It may involve discussion on a one-to-one basis with a person whose role is encouraging the development of teachers' professional skills. It may involve a personal scheme of systematic reading or research. It may, indeed, involve any combination of these. It may lead to the acquisition of some professional qualification, it may be undertaken with a view to

securing a particular type of appointment, or there may be no expectation of financial or academic reward. It may imply voluntary involvement or compulsion.

(Henderson 1978, p. 13)

In the BP project, the direct monitoring has been concentrated on in-service *courses* in primary mathematics. However, there is a good deal of other, often school-based, in-service work in primary mathematics. Much of this takes the form of work within an individual school, which can well be described as a 'joint problem-solving exercise' with a group of teachers from the same school studying the mathematics curriculum of the school and its teaching. Another factor that has begun to contribute to INSET in primary mathematics is the increased provision of posts of responsibility for mathematics in primary schools. In the mid-1970s, a few posts of this type were already in existence, but the publication of the HMI Primary Survey (DES 1978) and the Cockcroft Report (DES 1982) gave considerable impetus for the appointment of *mathematics co-ordinators* in many primary schools.

> The effectiveness of the mathematics teaching in a primary school can be considerably enhanced if one teacher is given responsibility for the planning, co-ordination and oversight of work in mathematics throughout the school. We shall refer to such a teacher as the 'mathematics co-ordinator'.
> ... We believe that every effort should be made to train and appoint suitably qualified teachers in as many schools as possible.
>
> (DES 1982, §§354, 358)

These mathematics co-ordinators have an important in-service role to fulfil with their colleagues, and are also in need of in-service support themselves in developing their role. The present study has thrown much indirect light on the role of the co-ordinator and on its problems, and on school-based INSET in general. These areas are discussed further in Chapter 10.

1.4 Methodology

The purpose of the project was to try to find types of in-service courses in primary mathematics, and ways in which these courses could be mounted and taught, so that the courses were effective in encouraging teachers to develop their mathematics teaching. It was thought that such development, if it occurred, would be unlikely to spring only from the overall structure of a course, but rather from details of the approach used in the course and from interactions between the participants, which it would be difficult to quantify. These details would only reveal themselves through extended observation and through personal discussion with the participants.

The monitoring of INSET in primary mathematics

For a number of years, much of the discussion of educational evaluation has concentrated on issues related to the development of appropriate methodology. The volume of readings entitled *Beyond the Numbers Game*, (Hamilton et al. 1977) has chronicled the gradual progression from the use of a quantitative methodology to a qualitative illuminative style of work. This style was characterised by Partlett and Hamilton as follows:

> At the outset, the researcher is concerned to familiarise himself thoroughly with the day-to-day reality of the setting or settings he is studying ... he makes no attempt to manipulate, control or eliminate situational variables, but takes as given the complex scene he encounters. His chief task is to unravel it; isolate its significant features; delineate cycles of cause and effect; and comprehend relationships between beliefs and practices and between organisational practices and the responses of individuals.
>
> (Hamilton et al. 1977, p. 14)

This seemed to be a clear description of the task which this project had set itself in the complex field of in-service courses in primary mathematics. Although illuminative qualitative methodology has most often been used in research related to the evaluation of curriculum development, this methodology also has important implications for all studies concerned with the formation of value judgements in education. Recently, it has developed into a variety of observational approaches to the study of classroom life (see, for example, Armstrong (1980) and Galton, Simon and Croll (1980)).

Hence, a qualitative methodology was chosen for the project's study. First, questionnaires were sent to all LEAs in England, and to former colleges of education, to determine the nature and extent of their provision for INSET in primary mathematics. This survey made it possible to classify the course provision into a number of general categories (see Chapter 2). Examples of events in each category were then selected for a close study, involving participant observation. In order to minimise the considerable travel problem, the events selected were largely confined to areas within reasonable travelling distance of Cambridge; these included East Anglia, parts of the Midlands and the outer London area. However, there was no concentration on courses within the Cambridge area itself; most courses monitored were in areas where the natural focus for INSET was a centre other than Cambridge. Another element in the choice was that the range of courses which could be selected for monitoring was dependent on the provision of INSET by LEAs and colleges at the time when monitoring was possible. A choice had to be made, too, between participating in a variety of activities of different types or covering a narrower range and comparing similar events. It was decided to give priority to covering as many different types of in-service course as possible, although, on a few occasions, rather similar types of course were observed in two different settings.

Throughout this report, the term 'provider' is used to refer to the person or people who actually 'taught' the course, rather than to the organisation under whose auspices it was mounted. The project is most grateful for the willingness of all the LEAs and providers who were approached to allow their work to be observed; in no case did a course that had initially been selected for study have to be withdrawn because either the LEA or the provider was unwilling to participate in the project.

The structure of primary schooling in the LEAs where courses were studied was representative of the main types of structure found in England. In some LEAs, there were infant schools (5-7) and junior schools (7-11), and some all-through primary schools (5-11). In other LEAs, teachers worked in first schools (5-8 or 5-9) or in middle schools (8-12 or 9-13). Nearly all of the teachers who participated in the INSET monitored were generalist class teachers, but a small number of specialist mathematics teachers in middle schools were included. Discussions with these teachers have only been quoted in two case studies, those of the Strategies and Procedures of Mathematical Problem Solving project and Open University courses and related INSET (Chapter 6); the activities described in these two case studies are rather different from those of the other studies. In all the other studies, the teachers who contributed to the case studies were generalist class teachers.

Soon after this project started, the Department of Education and Science decided to sponsor an evaluation of the *Diploma in Mathematical Education* of the Mathematical Association (Melrose 1982). This was the only other project which was studying in-service education in primary mathematics in England at that time, and so, as the field under study was so large, it was decided not to duplicate the efforts of others, and no courses for the Mathematical Association Diploma were monitored. MA Diploma courses are all situated in colleges (and their outposts) which also provide initial training for primary teachers. Consequently, in-service activity under the auspices of the former colleges of education is probably underrepresented in this study, although two college INSET courses of other types were studied in detail. It is impossible, however, to draw a hard-and-fast line, in terms of staffing, between college courses and others, because it is common for members of college staffs to contribute, at least to some extent, to LEA or Teachers' Centre courses, and vice versa.

When suitable courses had been selected, a case study was made of each. The method of investigation adopted was participant observation, combined with interviewing and the analysis of documents. The observer attended all the sessions of the shorter courses and a representative sample of sessions of the longer courses. Normally, both providers and course members knew the purpose of the visits. Whenever possible, providers were given an opportunity to discuss their aims for the course, sometimes before and sometimes after the event. In most cases nearly all of the participants were visited in their own schools after the end of the course; here it was possible to meet and ask questions of their head teachers and to see their classrooms, as well as to discuss the course with the teachers themselves. In most cases, the course advertisement and the handouts used during the course were available for study. In two cases, course members and their Heads responded to questionnaires. Details of the method of monitoring used are given in each case study.

The fieldwork of the project was carried out entirely by Sister Timothy, who selected the sample of courses to be monitored on the basis of the LEA and college questionnaires, and attended the courses as a participant observer. She also engaged in discussions with providers, and visisted the participants in their schools. It was intended that the project's report would also be written by Sister Timothy, but the volume of paper generated by the project was great, and the analysis took longer than had been anticipated. The observer made detailed notes during course sessions, and most of the interviews were taped and transcribed. This bulk of paper was analysed and case studies were written up from it. During the writing of the case studies, a number of recurrent themes emerged; these are discussed in Chapters 9 to 12. At the end of the project's time-span, the case studies were not completed. Pressure of work, both in a new appointment and in her own Order, prevented Sister Timothy from devoting as much time to completing the report as she would have liked. Consequently, this report is the joint work of Sister Timothy and of Hilary Shuard, who was involved throughout the project in its planning and in day-to-day discussion of the work with Sister Timothy, although she did not carry out any of the fieldwork of monitoring.

1.5 The 'effectiveness' of in-service education

It was no part of the intention of the project to pass judgement on particular courses of in-service education in primary mathematics, but it was impossible not to wonder how far, and in what ways, the courses monitored were 'effective'. The course members had no such scruples about commenting on the quality of the courses in which they participated; they frequently said to the researcher 'It was a good course', or 'I was a bit disappointed'. Sometimes, different participants passed very different judgements about the same course, but sometimes all those interviewed expressed rather similar sentiments. Consequently, it was important to try to spell out the notion of 'effectiveness' as it applies to in-service education in primary mathematics.

Is an in-service course effective if it achieves the goals that have been explicitly stated? Hamilton *et al.* (1977) have described the growth of qualitative styles of educational evaluation. Goal-free evaluators believe that if an evaluation of an educational event focuses only on the achievement of stated goals, it may miss much else that is of importance to the participants. It seems likely that, in in-service education, some unplanned outcome of a course may be as important for the participants as the planned goals. For an individual teacher, an apparent by-product of course attendance, such as a meeting with a stimulating colleague or a discussion over a meal, may be as important as what was planned by the course organisers. Therefore, both planned and unplanned features of the courses monitored are described in the case studies. It was important, however, to elucidate what the goals of the providers were, and to discover not only whether those goals were achieved, but whether the stated goals matched the needs and expectations of the participants.

A different aspect of 'effectiveness' is whether, and in what ways, in-service education produces 'more effective' teachers. This enquiry covers a range of questions.

What is 'effective' teaching?
What are the qualities of an effective teacher?
Can these qualities be developed through in-service education?

In many of the courses observed, the providers had some clearly defined overt goals, such as helping the participants to examine and understand the development of a mathematical topic throughout the primary years. Often, however, the fundamental goal of the providers was to communicate to the participants some aspects of their own understanding of what constituted 'effective' primary mathematics teaching, and their knowledge of how to carry it out. Detailed discussions with participating teachers were necessary in order to try to elucidate both whether a course was effective in achieving the providers' overt goals, and whether it was effective in contributing to a teacher's professional development (whatever may be meant by that phrase). Discussions with teachers were usually held some time after the end of a course, to allow time for reflection and for the teacher to put into practice any changes in teaching that the course might have initiated; by this time, also, the immediate impact of the course might have faded, and it might be possible to assess the extent of 'fade'.

1.6 A model of professional development

Gradually, by considering the many interviews with teachers which the researcher undertook, by thinking about their varied reactions, contributions and responses during INSET, and by reflecting on the experiences of teachers who were helping students on teaching practice, the researcher came to recognise an emerging pattern of professional development: this has finally been formulated as a four-stage model. In this section, this model is briefly outlined, so that the reader may bear it in mind when reading the case studies; it is further developed later in the report, on the basis of the description of teachers' INSET work given in the case studies. As in other models of development, the stages tend to recur when the individual encounters a new situation; hence the model seems to apply to the growth of individual teachers in the various roles they perform in the profession: as class teachers, as subject teachers, as co-ordinators, and as administrators.

Four stages are highlighted in the model, not because they are discrete phases, but because the characteristics listed for each stage tend to show themselves as a teacher becomes more experienced and skilled in a particular situation or type of work; these stages seem to recur as the teacher moves on to new roles. The stages are:

1 *Initiation:* At this first stage, teachers lack experience of the situations in which they find themselves; they seek *information* about *what* to do and *how* to do it.

2 *Consolidation:* Teachers at this stage achieve a synthesis of action through repeated experience and some experimentation within the situation; they establish a relationship at a functional level between what needs to be done and how to do it; they develop *teaching skills* which evolve into adequate although limited *personal teaching styles*.

3 *Integration:* By this stage, security and confidence have grown as a result of the teacher's competence; as long as this does not merely lead to complacency, it provides a basis for further development. Teachers at this stage become aware of what others do and how they do it, and relate their work to a consideration of other age-groups and other subject-areas; all these considerations contribute to the development of more *flexible* and *adaptable* teaching styles. Teachers are now able to *share* their experiences freely with others, unhampered by fear. Their own questioning now relates to broader issues, focusing on curriculum development and on individual learning needs.

4 *Reflection:* Full development is now demonstrated by teachers' explicit personal awareness of the values that determine their decisions, although they continue to make many professional decisions intuitively, on the basis of accumulated experience. In this stage, several changes of outlook are synthesised: teachers make a more careful study of how *learning* occurs, and this justifies and modifies their teaching methods. They give further thought to what is taught, so that the content of the curriculum becomes a servant of the

process. Curriculum integration becomes more than merely an awareness of continuity and progression as the child grows older, or of interrelations within the content; the curriculum now becomes the servant of *education*. Teachers' own enquiries start by asking 'Why teach this or that topic?'; the enquiry develops so that they grapple with questions such as 'Why teach mathematics?'. At this stage, a teacher's personal teaching style develops into a personal *educational philosophy*.

When interviewing teachers, the researcher met teachers in all these stages of development, in their various roles as class teachers, as co-ordinators and as Heads. In-service providers also pass through the same stages; a teacher who is asked to contribute to an INSET course for the first time, or a new college lecturer, certainly needs to seek information about what to do and how to do it; skills in the teaching of adults only gradually develop — even though the teacher may well already have reached the fourth stage (reflection) in working as a class teacher in a primary school.

A rather similar developmental model has been put forward by Watts (1981). She uses only three stages:

- survival
- a middle stage
- mastery

However, her descriptions of her stages, and her quotations from conversations with teachers, involve very similar points to those described in our model.

CHAPTER 2

Sources and types of INSET provision in primary mathematics

2.1 Sources of provision

Taylor (1978) listed several types of institution that carry responsibility for INSET, and classified the major types of INSET undertaken by each. The following list is adapted and shortened from that given by Taylor:

Institution	INSET work undertaken
School	Training and support for staff development.
Teachers' Centre	Short courses; longer courses where needed. Informal working groups. Demonstration of materials.
LEA	Stimulation, support and provision of INSET. Administration and liaison.
College/Dept of Education in public sector	Award-bearing courses; other courses of substantial content. Conferences and short courses.
University School of Education	Award-bearing courses. Research and enquiry.

In the detailed study made by the project of the provision of INSET in primary mathematics, the focus is on the variety of provision of *courses* by the major organising bodies; these bodies are the LEA Mathematics Advisory Team, local Teachers' Centres and the former colleges of education, which now have a substantial commitment to in-service education as well as to initial training for teaching.*

*In this report the former colleges of education will be referred to as *colleges*, although some of them are now departments of Polytechnics and others have become colleges of higher education. Their major common characteristic of importance in the study of INSET is that their courses include initial training for primary teaching as well as INSET.

Sources and types of INSET provision

In early 1980, postal questionnaires were sent to all LEAs and to all colleges in England, requesting them to give details of their INSET provision in primary mathematics in the 1979—80 session. The response rate to the questionnaire to LEAs was only about 12 per cent and several LEA Advisers complained about the number of similar enquiries they now received:

> I now receive so many of these documents that the demand on my time to complete them fully and seriously would prejudice the service I offer to our own teachers. I am sorry to be so unhelpful.

The response to the questionnaire to colleges was rather better (44 per cent), perhaps because the enquirer was well-known to many of those who completed the questionnaires.

In the end, the replies received, with the considerable number of Teachers' Centre programmes and advertisements for courses that the researcher was able to collect, enabled a rough classification of the variety of provision to be made. This is shown in Table 2.1; it largely confirms and supplements for primary mathematics the pattern of general provision given by Taylor.

In Table 2.1, as well as listing the *strucures* and *organising bodies* of the courses, a sample of the different types of *content* offered in courses of these types is shown; there is also an indication of some of the *goals* which the courses seemed to have. This last column of the table was completed from course descriptions and advertising material, and after observation of sample courses of the types shown.

Long courses, both full-time and part-time, are entirely provided by institutions of higher education, especially the colleges. There is a small amount of provision of long courses in primary mathematics in the university sector; for instance, a very few primary teachers take the Cambridge M.Phil. in mathematical education. The project did not enquire into provision in the university sector, except that part of the Open University's provision in primary mathematics was monitored. The reasons for this concentration of provision are that colleges are able to provide award-bearing courses, and the patterns of their organisation enable them to make the long-term commitments needed, both in lecturing staff and in accommodation. Another factor is that the Mathematical Association only licenses courses for its Diploma in Mathematical Education in institutions that also provide initial training.

LEAs are usually the instigators of short-term provision. Their close contact with teachers, through Teachers' Centre panels and through the Advisers' visits to schools, makes them aware of teachers' expressed needs and able to respond to these, as well as to immediate local and national initiatives.

The venues for short courses and working group meetings depend on the geographical nature of the locality, particularly whether it is

Sources and types of INSET provision

Table 2.1 Types of provision for INSET in primary mathematics

Clasification	Structures	Organisers	Content (*sample*)	Some goals
Long full-time courses	One year One term	Colleges [Universities]	Mathematics Maths curriculum Maths education	Extended time to consider classroom issues Practice to theory
Long part-time courses	1, 2 or 3 years: evenings and/or ½ day per week	Colleges Open University [Universities]	Mathematics Mathematics education	Related to Diploma Theory Theory to practice
Short courses	One session Weekly sessions One day 3–4 days	LEAs Teachers' Centres Colleges	Current issues Schemes Apparatus Assessment Language	Information Understanding [Often undefined]
Working groups	Occasional meetings in series	Associations LEAs Teachers	Guidelines Classroom research Local issues Topics Apparatus	Developing progression Deepening understanding of learning and teaching Follow-on to other courses
Residential	Weekend 2–3 days One week	LEAs Colleges DES Associations	Theme — a recent report Infant or junior maths Co-ordinator's role	Opportunity to share ideas, to stand back and reflect

urban or rural. Teachers' Centres, both multi-purpose centres and specialist Mathematics Centres*, form a natural focus for INSET in primary mathematics. Weindling, Reid and Davis (1983) estimate that in 1979 there were about 485 Teachers' Centres in England and Wales, of which 385 were multi-purpose. It was not stated how many of the 100 specialist centres were mathematics centres; the Cockcroft Report (DES 1982) suggests that these were few in number. Perhaps 400 Teachers' Centres provide short courses in primary mathematics, so a very substantial part of the whole provision of INSET in primary

*In this report, *Teachers' Centre* will denote either a multi-purpose Centre or one of the small number of specialist Mathematics Centres.

mathematics must be based in these centres. The provision of short residential courses is constrained by the availability of suitable accommodation; these courses are usually held in LEA Residential Centres, although a few are mounted in colleges during vacations. The project did not study how working groups in primary mathematics operate, but they form a considerable part of the activity of Teachers' Centres. Weindling *et al.* suggest that more teachers use centres for meetings of working groups than for courses.

2.2 Types of provision in Teachers' Centres

Because of the importance of the provision of INSET in primary mathematics in Teachers' Centres, a representative list of activities in primary mathematics offered in Teachers' Centres during the period 1979—82 has been compiled from the many Centre programmes received by the project. This list is intended to demonstrate the variety of types of course that are found in Teachers' Centres, and the rapidity with which a Centre can respond to local and national issues. No single Centre would provide the range of activities shown, but they were all offered somewhere, and many centres do offer a great range of courses. The list covers the three-year period 1979—82 during which the project operated, because centres do not repeat the same programme every year, but often seem to run on about a three-year cycle; 1979—82 was also a period in which several national reports discussed primary mathematics, and when several new textbook schemes were published, and we can see how the Centres responded to them. Titles, times and brief descriptions of the courses are given; these are usually abbreviated versions of the course publicity leaflet, or are taken from termly brochures. The names of the course providers are omitted. Most Teachers' Centre activities take place after school, but there are some day courses, usually mounted by the LEA Advisory staff and involving release from school. A few Centres provide activities on Saturdays or sponsor courses at Residential Centres.

During this period, a growing trend (which Teachers' Centres were reflecting) was the development of interest in microcomputers in the primary school. Many Teachers' Centres offered courses in this area under the heading of mathematics, but the project decided to regard computing as a separate area, and it is not included here.

INSET courses in primary mathematics at Teachers' Centres
(a composite list compiled from several centre programmes)

Autumn Term 1979

Teaching number bases in middle schools: two sessions, 16.45–18.15. During the first session, the reasons for teaching number bases will be critically

Sources and types of INSET provision

reviewed; the second session will be concerned with ways of improving the teaching of number bases.

The language of maths: a practical workshop: one session, 16.30–18.30, and continuing termly sessions. This small research group is continuing its investigation into language problems experienced by children in their learning of mathematics. New members are welcome.

Mathematics co-ordinators in middle schools: one session, 16.30–18.30. A follow-up to the residential course held last term.

Mathematics and the HMI Primary Survey: one session, 16.30–18.00. The implications of the survey for both class teachers and head teachers will be examined.

Mathematics and the APU: one session, 16.30–18.00. The APU is engaged in monitoring the type of mathematical thinking which is appropriate for children in different stages of development. The talk will be of interest to all teachers of mathematics, primary and secondary.

Diploma in Mathematical Education: weekly meetings for two years, 14.00–19.00. A course leading to the award of the MA Diploma, run by college tutors.

Mathematics apparatus in action: two sessions, 16.30–18.30. The first session will discuss possibilities offered by two items: nailboards and pegboards. The second session will deal with apparatus requested by teachers at the first session.

Exhibition of pupils' mathematics: in the Central Library for two weeks. It is intended to give the general public an idea of what happens in mathematics teaching from the nursery classroom to A-level. Teachers will also find the exhibition interesting.

Spring term 1980

Sample lessons in mathematics: five sessions, 19.00—21.00. A series of demonstration lessons on mathematical topics, which will be of particular interest to non-specialists who are teaching mathematics in middle schools and to specialists who are looking for fresh approaches.

Mathematics for the primary classroom: one session, 9.15–16.30. A day conference for all mathematics co-ordinators. This day is run by the Mathematics Advisory Team to raise a number of important issues including the HMI Primary Survey and Mathematics 5–11. This will be the only day for mathematics co-ordinators this year, and it is urged that they make every effort to come.

Teaching mathematics from Fletcher: five sessions, 16.30–18.00 at Sycamore Lane Junior School. This course was requested by the staff of the school, but other teachers are welcome. Topics covered will be school organisation, the language of sets, the sequential development of computation and problems of tailoring the scheme to children of different abilities.

Development of early mathematical concepts: five sessions, 16.15—18.30. This course is open to teachers and nursery nurses. Four mathematical themes will be taken from the Schools Council project *Early Mathematical*

Sources and types of INSET provision

Experiences. Practical workshop sessions will include discussion of teachers' observations of children using mathematics equipment. Topics: space and shape, comparison, towards number, the passage of time, record keeping.

Summer term 1980

Nuffield Mathematics 1 and 2: one session, 16.30–18.00. A demonstration of the new Nuffield Mathematics project materials for children up to 7–8 years.

Concepts in mathematics: one session, 16.30–18.00. A talk about the work of the CSMS project with reference to the team's research in the middle and secondary years of schooling and to the APU first primary survey.

Demonstration lesson: one session, 16.30–18.00. A well-known mathematics educator will give a demonstration lesson with a group of able children of 9–11 who are particularly interested in mathematics. The lesson will last one hour, with discussion for half an hour.

Aspects of teaching: seven sessions each term, 16.30–18.30. For newly and recently qualified first school teachers. The eighth unit, lasting four sessions, is on aspects of mathematics:
1 Beginnings: the notions that a child must absorb before beginning formal number work; apparatus and classroom activities.
2 Shape and spatial relationships.
3 Beginnings continued.
4 An opportunity to see a range of children's work and to discuss methods of organisation.

Autumn term 1980

Creativity in mathematics: one session, 19.00–20.30. If we can recognise and assess creativity in mathematics in our pupils we can reward and foster it. The lecture will look at recent attempts to identify aspects of creativity in children of about 11, and consider the relevance of personality traits such as a willingess to take risks in mathematics. The audience will be invited to have a go at some tasks which require divergent thinking in a mathematical context.

Calculators in action: one session, 16.30–18.30. The meeting will have two main aims: to introduce in a workshop session some activities with calculators, and to discuss the issues that arise from such activities in the classroom. The two speakers have worked in the reality of classrooms as they are, and are acutely aware of the concern teachers have about calculators.

The development of logic in infant and junior schools: Saturday, 9.30–16.00. A one-day conference to examine the development of logic and reasoning, and its teaching, during the whole of the primary age-range.

Organising the classroom for mathematics: one session, 16.30–18.30. Few teachers find easy solutions to the problems of translating what they believe to be important principles of learning into the reality of everyday classroom work. This will be the first of a series of working meetings; it will establish concerns and decide on ways of approach; a video will be used to start discussion.

Guidelines in mathematics: one session, 17.00–18.30. An initial discussion on writing Guidelines for teaching mathematics in the 5–11 age-range. If sufficient teachers are interested, a group will be formed to write Guidelines for use in the teacher's own school. It is anticipated that this will be a long project, taking one or two years.

Revision course on mathematics: ten sessions, 16.30–17.30. To help middle school non-specialists to improve their own mathematics knowledge.

Spring term 1981

Omnigraphics: one session, 17.00–18.30. Some interesting ideas about the visual manifestation of mathematical concepts. A demonstration and exchange of views.

Primary mathematics assessment: one session, 17.30–19.00. A pilot scheme will shortly be introduced involving the use of a profile of mathematics skills. The scheme is to be tried with a view to adopting the profile as a guidance test for use in the 9–13 age-range in future years. The profile will be available for inspection, and those present will be given every opportunity to express their own views on the testing of mathematics in the primary school.

Working in groups in mathematics: one session, 16.30–18.30. Teaching mathematics through working with groups of children; criteria by which teachers can plan for and judge the effectiveness of managing children's learning of mathematics in group activities will be offered.

Mathematics: organising your classroom: one session, 16.30—18.30. A follow-up to 'Working in groups in mathematics'; further seminars will be arranged if there is a demand.

Games in mathematics: one session, 16.30–18.30. A workshop session led by a local primary teacher; there will be an opportunity to play and discuss mathematical games.

Follow-up to games session: one session, 16.30–18.30. Teachers were very impressed with the materials shown at the games session, and with the method of working, but expressed doubt about the practicality of classroom organisation. The speaker has now agreed to do a demonstration lesson with local children.

Residential weekend: Creative mathematics in play. This course will explore young children's activities, and identify mathematical experiences within them. The development of mathematical concepts in young children, and exploiting children's curiosity and capacity for enjoyment, will also be studied.

Summer term 1981

Mathematics co-ordinators' meeting: one session, 13.30–17.30. In the past 18 months, three major documents have been published which relate to the work of the mathematics co-ordinator. This conference is based on the requests of co-ordinators for information and the opportunity to explore some of the views expressed in these documents.

Maths film preview for primary teachers: one session, 19.30–21.30. An opportunity to see what the Centre holds, and notes on how to borrow.

Sources and types of INSET provision

Developing your group work: one session, 9.00–16.30. An intensive day for teachers who have experience of group work in mathematics, and who wish to develop their practice.

Peak Mathematics dissemination meeting: one session, 16.30–18.00. A teacher and a representative of the publisher will discuss features of this new scheme. Materials will be available for inspection.

Autumn term 1981

Some uses of calculators in primary and middle schools: one session, 16.30–18.00. Basic calculations using simple calculators; there will be no advanced scientific or programming work. Teachers are asked to bring their own calculator if possible.

Middle school mathematics co-ordinators: one session, 14.00–16.00. Some thoughts on the forthcoming Cockcroft Report and an opportunity to discuss matters raised by members. This meeting is on the same day as the Taskmaster display, which co-ordinators may wish to inspect.

Evaluating mathematics materials: three sessions, 13.30–16.30. The speaker will suggest a number of questions, the use of which will help the teacher to assess the relevance of mathematics materials to what and how they want children to learn.

Mathematics in the early years: three sessions, 13.30–16.30. Developing the concepts of shape, capacity, weight and length by using apparatus. By doing and discussion, participants will develop an understanding of the maths concepts behind the activities. A chance to report back on what you find during the week, and to bring along materials of whose potential or use you are uncertain.

Spring term 1982

The Cockcroft Report: one session, 14.00–16.00 for middle school co-ordinators, 16.30–18.00 for primary teachers. A member of the Committee will give a preliminary consideration to the Report and its implications for schools. Note that it may be necessary to rearrange this meeting if publication of the Report is delayed.

The use and application of mathematics materials in schools: six sessions, 16.30–18.30.
1 What's at the back of your cupboard?
2 From Unifix to calculators.
3 An overview of materials 5–13.
4 Materials in the infant classroom (two teachers will speak).
5 A junior classroom (a teacher will speak).
6 How do you work out priorities?

Summer term 1982

The Cockcroft Report: further implications for primary schools: one session, 16.30–18.00. A member of the Committee will continue his commentary on the Report. This session will not deal with calculators, which will be the subject of the next session.

Are you expected to teach area?: one session, 18.45–20.00. Where do you

Sources and types of INSET provision

start? How far do you go? What resources do you need? Hear and see some ideas from practising teachers; examine relevant materials; discuss problems and ideas.

In reading this list, we cannot help noticing the variety of provision. Probably in any local area where a teacher is seeking help about primary mathematics, there will be sessions at the Teachers' Centre to introduce newly published primary mathematics schemes; 'user groups' often spread around the expertise of teachers who use established schemes, and there are usually sessions on the use of mathematical apparatus of various types. After that, the variation is enormous and the courses offered depend greatly on local talent. Usually, some meetings are inspired by the Mathematics Adviser, and are used for the development and spread of LEA policy on mathematics; meetings of co-ordinators, the discussion of testing and guidelines, and the dissemination of the Cockcroft Report are of this type. Other meetings are arrranged by the Centre's mathematics planning group, or some similar group of teachers, or stem from the Teachers' Centre Warden's interests and contacts. One Centre became interested in developing group work in primary mathematics, and ran a concentrated series of meetings on this; no other Centre whose publicity was received by the project displayed a similar interest. Mathematical games and puzzles are a common attraction, but only one Centre knew someone who was able to run a course on the mathematics inherent in young children's play. One Centre was visited by several mathematics HMIs during the period; another relied on attracting academics in the field of mathematical education.

The response of the planning group to local and national needs is sometimes a question of 'What can our local teachers offer in the way of in-service?' and 'Who else might we persuade to come?' The grapevine flourishes, and people who run a successful course at one venue in an LEA are likely to find themselves in demand at a number of adjacent Teachers' Centres. Centres undoubtedly provide an interesting diet in primary mathematics, even if it is sometimes a rather haphazard one. In very few of the courses at the Centres was any serious study expected; very occasionally, reading was expected before a course session, and a few rural Centres were used as outposts of higher education institutions, enabling award-bearing courses to be brought to isolated groups of teachers. However, in all Centres, many sessions contained practical activity and an opportunity for discussion with other teachers, and no doubt brought together teachers with similar interests.

2.3 Types of INSET provision in colleges

A questionnaire was sent to all colleges in England, early in 1980, asking for details of all their in-service provision in primary mathematics in 1979—80. Recent amalgamations of former colleges of education into larger institutions made it difficult to know who in the college should receive the questionnaire, because in some colleges members of staff who might be expected to provide INSET in primary mathematics had become members of mathematics departments; in other colleges they were placed in education departments while mathematics departments were only concerned with academic mathematics. In the end, the questionnaire was sent to the INSET co-ordinator of the college, in the hope of obtaining a global view. The response rate was 33 out of 75 questionnaires sent out, or 44 per cent.

All the 33 colleges that replied did some in-service work in primary mathematics, and again the pattern of provision was similar to that shown by Taylor (1978). Table 2.2 shows the numbers of colleges that provided different types of INSET in primary mathematics.

Table 2.2 gives a rather misleading impression of the proportion of the colleges' effort that was spent on different types of INSET and the numbers of teachers reached by it. A great deal of college effort was expended on the Mathematical Association Diploma, several colleges mounting courses at more than one centre, so that 24 separate courses were provided by the 19 colleges. Moreover, Diploma groups were usually large, having an average of 17 teachers on each course. Including both first- and second-year Diploma students, there were 569 teachers following MA Diploma courses in the 19 colleges in 1979-80. The in-service BEd made a smaller contribution to INSET in primary mathematics. The groups were smaller, consisting of only ten teachers on average, and the time devoted to mathematics was less; commonly, the mathematics was a component of a Primary BEd,

Table 2.2 Provision of INSET by colleges (1979—80)

Type of provision	No. of colleges	Percentage
MA Diploma*	19	58%
In-service BEd (maths option)	9	27%
Diploma/Adv. Dip. (maths option)	6	18%
1-term full-time	3	9%
1-year part-time	2	6%
Short courses	20	61%
School-based consultancy	19	58%

*Similar courses validated by bodies such as the CNAA are included here.

Sources and types of INSET provision

and occupied a unit of perhaps 50 hours, while the length of an MA Diploma course was at least 200 hours.

The major contribution to Advanced Diploma courses in primary mathematics was made by one college, which had three very large groups, averaging 27 teachers. Other Diploma and Advanced Diploma courses made a negligible contribution to the numbers of teachers on substantial courses, and there was a total of only 35 teachers on the three one-term full-time courses in the colleges. Thus, it is clear that, in 1979–80, the MA Diploma (together with similar courses) was the major contribution of the colleges to substantial INSET in primary mathematics.

We now consider courses less substantial than Diploma courses. In no case does the contribution of a college to short courses seem to have been as great as a Teachers' Centre might be expected to provide during a year. One college ran six short courses, but the average provided by each of the 20 colleges was 2.6 courses. However, many of these courses were rather more substantial than the usual Teachers' Centre course, and often ran for ten meetings or more.

A particular feature of the work of some colleges was a commitment to school-based work. Nineteen colleges took part in school-based INSET, providing courses for groups of schools and individual schools, consultancy and discussion. Again, the numbers were small, 36 'events' being organised by the 19 colleges, although some of these 'events' were long-running series of consultations. In the case of school-based INSET, there seems to be some difference of policy between polytechnics and other colleges: only three of the eleven polytechnics which responded provided this service, while 16 of the 22 other colleges did so. The initiative for school-based work was usually taken by either an Adviser or the Head, and these two people approached colleges in nearly equal numbers.

Colleges were also asked which types of INSET they regarded as most important among those that they might provide. They were asked to give three choices, in order of priority, with reasons. Colleges naturally chose largely those types of INSET of which they had experience, and their preference was very clear; the MA Diploma and school-based work tied at the top of the list, long full-time courses (one-term or one-year) came next, substantial part-time courses (one session per week for a term or a year) had some support, and short courses were not favoured.

2.4 Reasons given by colleges for their INSET preferences

The extracts from comments in the questionnaire replies, listed below, typify the reasons college tutors gave for their priorities in providing INSET in primary mathematics.

Sources and types of INSET provision

MA Diploma and similar courses

Trains teachers as maths co-ordinators and therefore indirectly affects other teachers and types of school curricula.

By helping such teachers we are indirectly helping many other teachers, whom, we hope, those on the course will help.

There is real need for more 'specialist' teachers of mathematics in primary schools — at least one in each school.

The primary/middle age-range is the area most in need of 'leaders' in the subject.

All primary schools need a person with expertise in maths to lead and advise other teachers.

Change is never instant; it takes time to influence teachers, give background and develop thinking.

These courses give teachers something to aim for; they help to improve the status of primary teachers and of the subject in primary schools.

There is enormous demand for this from teachers in the locality.

A minority interest for those teachers who have a previous good background in mathematics.

School-based work

By concentrating on one or two school staff over a good period of time there may be some permanent influence.

Although school-based work is uneconomic, one meets *all* the teachers in the school, not merely the keen ones, and also those who are not free from home ties to travel to courses. When teachers realise one is able to contribute in a classroom they appreciate any advice given.

An attempt to involve all staff — not only the regular course-goers. We have also run discussions for all the feeder schools of a secondary school. This allows access to the reluctant school, and has proved to be very productive.

Important for the staff as a whole to reflect and co-operate.

Particularly valuable where teachers feel isolated from outside contacts and help in mathematics, and where teachers lack confidence in teaching the subject.

There is a need for back-up in schools, as so many teachers and heads lack confidence in their own ability to organise mathematics. The requests usually follow courses where contact has been established.

This provides the opportunity to consider the mathematical problems or strengths of an individual school. It is less generalised than a college course.

Many teachers are insecure and like help within the school. Also more teachers are involved and such courses tend to have a more lasting effect.

This provides the opportunity for real dialogue between school and college. The school benefits from the resources of the college and college staff have an opportunity to work with children.

Sources and types of INSET provision

Provides two-way benefit: helps schools and keeps lecturers in schools.

This has significant effects on the work being done in the school and is also of immense value to the lecturer involved.

The demand and need is there.

This is what teachers ask for, but tutor workloads do not at present allow us to do it.

Full-time one-term or one-year courses

This course would be of greater benefit to teachers because it would remove the intensity of the day-to-day pressures of school.

The teacher is released from other responsibilities and can concentrate in depth on the course and its requirements.

Gives teachers the opportunity to think about what they are doing away from the classroom situation.

This allows the person responsible for maths the opportunity to rethink schemes of work, plan staff development, etc., without the problems of teaching but with resources such as equipment and major schemes of work available.

By devoting a whole term to such studies, teachers are able to analyse mathematical learning in depth and to develop considerable competence so that they can pass on advice to their colleagues in school.

Only by this kind of concentrated course can you hope to influence teachers in their approach to teaching mathematics.

A lengthy course is essential if the members are to know each other and make worthwhile contributions to discussion.

Substantial part-time courses (at least 10 meetings)

Content can vary with need, e.g. Guidelines, assessment, special topics, new books.

To familiarise teachers with aims and content of new school texts and materials.

Brings teachers from various schools together with mutual benefit; the resources of colleges are available where necessary — e.g. use of computers.

This is really responding to the needs of teachers and allowing them to use the resources of the college to work through specific problems in school.

Teachers on our Diploma course expressed the desire for courses which we had considered to be initial teacher training — e.g. methods of teaching fractions.

In this type of course, although concentrated study is not possible, teachers have the advantage of immediate return to the classroom to try out ideas discussed in the course.

On a one-day-per-week course, the teacher is released from classroom responsibilities and able to devote herself to the course — contrary to evening courses, when the teacher is worn out after a day's teaching. A day course promotes greater course spirit — enables interaction within the group. Also allows tutors to visit teachers in their classrooms.

The need is to get teachers out of the classroom to *do* mathematics together over a long period of time and to enable teachers to attend who, for various reasons, are unable to attend in evenings.

There is an evident need to support the work of overburdened LEA advisers by input at some depth.

There is a considerable demand for such courses. Being less demanding on lecturers' time it is possible to help more teachers, albeit in a more superficial way than by other types of activity.

Facilities are needed for self-help groups with support from a college lecturer: the health of mathematics teaching depends ultimately on teachers who want to work at the problems. Courses for reluctant teachers or those seeking easy solutions are unlikely to have much effect.

In these extracts, we can see tutors expressing anxiety to find effective ways of helping primary teachers to develop their mathematics teaching, but opinions differed about the best structures to achieve this end, and about how best to use scarce resources of time and manpower. Some tutors put more emphasis on improving teachers' own mathematical knowledge; others favoured work closely related to the classroom. Respondents were also well aware that, to provide effective INSET, they need to keep in close contact with classroom reality, and they valued school-based work for this reason, as well as for its role in bringing a whole staff together to work on developing the school's mathematics.

2.5 Discussion

It would seem from the results of the two enquiries described above that there is little overlap between the styles of INSET favoured by teachers' centres and colleges. Teachers' centres largely provide short courses on the basics of primary mathematics teaching — apparatus and schemes — and bring in outside speakers to discuss national reports and other work in mathematical education. The commitment involved in their courses is usually very short-term. Colleges, on the other hand, attract teachers who are prepared to devote themselves to a longer-term commitment and to a more academic type of study.

A third type of INSET provider is the LEA Mathematics Adviser. Many courses arranged by Advisers take place in Teachers' Centres, but some are short residential courses held in LEA Residential Centres. Many of the Adviser's courses are policy-related: conferences for co-ordinators, the writing and dissemination of Guidelines, assessment, transition from primary to secondary schools; other courses fill needs of which Advisers are aware: dissemination of the Cockcroft Report, encouragement to develop the use of calculators in primary schools, the needs of the extremes of ability, and so on.

It would not perhaps be too much of an over-generalisation to say that, in general, the Adviser provides INSET for the development of the LEA's mathematics teaching, rather than for the needs of individual teachers or particular schools. Individual teachers turn more to the Teachers' Centre or to a local college, and individual schools in need of long-term help with their mathematics teaching can find very little external help. A fortunate school might find a college tutor who had time to work with its staff over a period, but this type of help can only be available to a tiny minority of schools. Thus, the importance of the co-ordinator in developing the mathematics teaching of a whole school becomes even more evident.

CHAPTER 3

An experiment in monitoring — an LEA course for co-ordinators

3.1 Methodology

The first course to be monitored was an LEA course for mathematics co-ordinators. The methodology was experimental; a very interesting course was studied, and is reported here. However, the main objective was to develop a procedure for monitoring future courses. The researcher attended the course as a participant observer, and took as full notes as possible during the course sessions. Very soon after the end of the course, she sent questionnaires about the course to the co-ordinators who had attended the course; a different questionnaire was sent to the teacher's Head. The questionnaires to the co-ordinators asked how useful the participants had found each course session, and that to the Head enquired about how the co-ordinator had put the ideas of the course into action in the school.

These questionnaires were sent to eleven schools, representing nearly half the twenty-three teachers who attended the course. Seven of the schools were asked to return the completed questionnaire by post. It was expected that only some of the questionnaires would be returned; in fact, three questionnaires were returned. Although a return rate as low as 43 per cent is quite usual for a postal questionnaire, it was felt to be insufficient to enable the researcher to gain a reasonable range of impressions of the course. Hence, when other courses were monitored, postal questionnaires were avoided whenever possible, and personal interviews with participating teachers were used instead. It was found that a very much higher proportion of schools — up to about 90 per cent — were willing to receive a personal visit from the researcher, when she wrote to the head teacher to ask to visit the school and discuss the course. On this occasion, the researcher visited the remaining four of the eleven schools that had received the questionnaires, and 'talked through' the questionnaires with the teacher and with the Head. Although it was comparatively easy to gain access to teachers by offering to make a personal visit to the school, the method of 'talking through' a questionnaire with a teacher was felt to be too restrictive to elicit personal comments that would throw light on unexpected as well as expected aspects of the teacher's reaction to the course. In later interviews, the researcher used the technique of preparing the areas to be discussed in advance (see Henderson 1978). This method was usually found to produce a relaxed and informal discussion with the teacher, who seemed to produce spontaneous reactions to the course. Head teachers seemed to be equally frank about their impressions of what the teacher had gained from the course.

3.2 Structure of the course

This course was one of a series that was provided by an LEA for the mathematics co-ordinators in its primary schools. Some of the primary schools in this LEA were separate infant and junior schools, and others were all-through primary schools. The intention of the Mathematics Adviser was that, over a period of years, all the mathematics co-ordinators in the LEA's primary schools would attend a three-day course of the type that was monitored. This particular course was held at a Teachers' Centre on three Fridays in February and March. There was an interval of two weeks between successive days of the course, and twenty-three teachers and the researcher attended the course.

3.3 The first day

After an introduction, in which the Adviser emphasised that the purpose of the course was to examine the *role* of the co-ordinator, the first session concentrated on ensuring that the teachers were acquainted with three important reports on primary mathematics that had been published recently. These were the HMI Primary Survey (DES 1978), *Mathematics 5-11* (DES 1979) and the first Primary Survey of the APU (APU 1980); the Cockcroft Report had not yet been published. The Adviser encouraged the co-ordinators to recognise that the criticism of mathematics teaching contained in these reports called for more emphasis on understanding, discussion, mental as well as written facility with number, and that mathematics needed to be used in an everyday context.

An experiment in monitoring

In preparation for the course, the teachers had been sent a task to carry out in their own schools. They were asked to give a test of arithmetic processes built around the purchase of a bag or tube of sweets (Figure 3.1); the test was thought to be suitable for children of

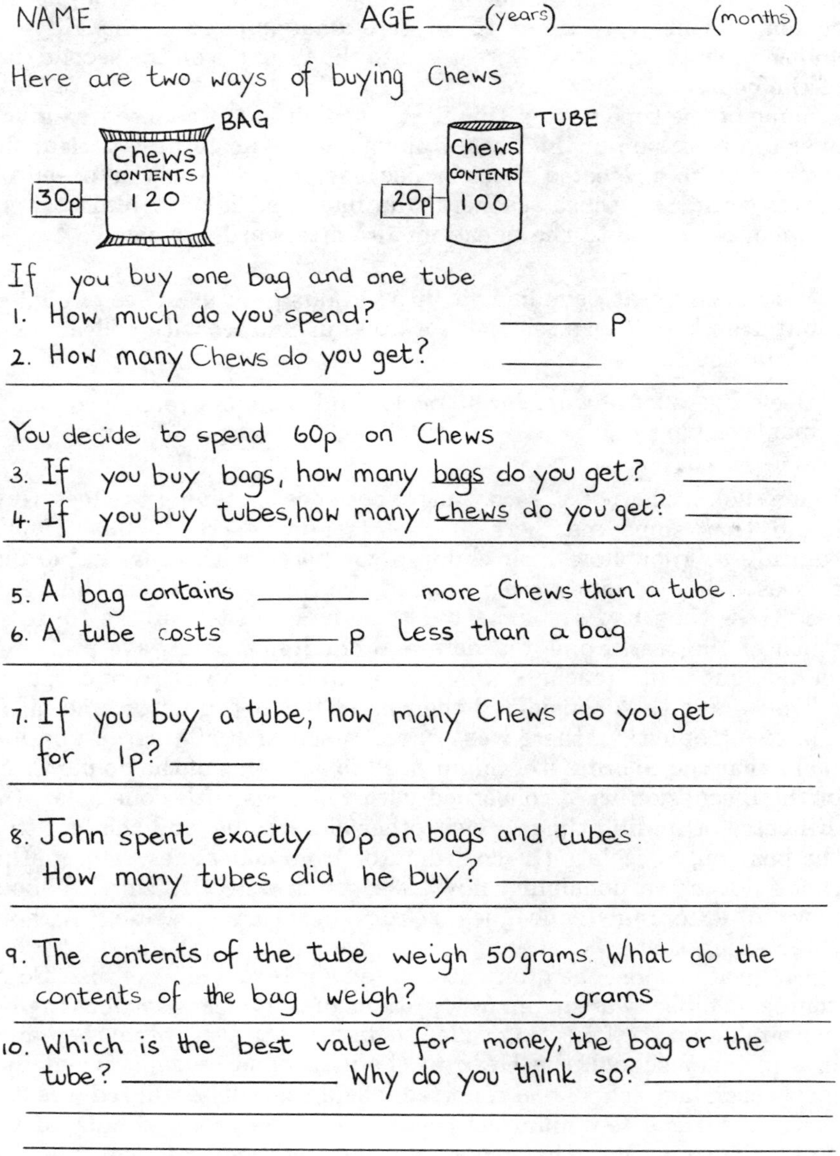

NAME _____ AGE _____ (years) _____ (months)

Here are two ways of buying Chews

BAG — Chews CONTENTS 30p 120

TUBE — Chews CONTENTS 20p 100

If you buy one bag and one tube
1. How much do you spend? _____ p
2. How many Chews do you get? _____

You decide to spend 60p on Chews
3. If you buy bags, how many <u>bags</u> do you get? _____
4. If you buy tubes, how many Chews do you get? _____

5. A bag contains _____ more Chews than a tube.
6. A tube costs _____ p less than a bag

7. If you buy a tube, how many Chews do you get for 1p? _____

8. John spent exactly 70p on bags and tubes. How many tubes did he buy? _____

9. The contents of the tube weigh 50 grams. What do the contents of the bag weigh? _____ grams

10. Which is the best value for money, the bag or the tube? _____ Why do you think so? _____

Figure 3.1 Test of arithmetic processes

all primary ages from top infants upwards to attempt, but it was certainly not expected that they would all complete it. This task was intended partly to ensure that the co-ordinators made contact with other teachers in their schools about the course they were going to do, and partly to focus the co-ordinators' attention on the range of attainment among the children in their schools. The teachers brought the school's results with them; these were amalgamated by the Adviser and the combined results were given to the teachers on the second day of the course.

Some of the time during this first morning of the course was given over to discussion of this test. Opinions of its usefulness varied; the infant teachers rejected the test because they said it was based on junior work, and those teachers who had used it with juniors were divided, as replies to the questionnaire afterwards revealed:

> Very useful — it gave insight into various practical uses of computation, and it was relevant for making contact with colleagues about the course.

> Useless — a lot of work for no profit, and it did not really provoke many talking points.

When the teachers discussed their experiences in giving the test, they found that some members of their staffs reacted competitively, wanting to know how their children compared with those in parallel classes, while others were pleasantly surprised at their children's results — they had not expected their children to be able to do so much of the test. Some teachers and children would have preferred 'sums', but other teachers were happy to move away from sums.

The course then considered the role of the co-ordinator, and, after a brief introduction, there was general discussion of a list of possible tasks that the co-ordinator might need to do. Some of the points made in the discussion were concerned with relations with colleagues, the difficulty of finding time to visit other classrooms, and the fact that the post might isolate the co-ordinator from colleagues: 'Oh, maths, that's not my responsibility now'. The group were also worried about how to encourage colleagues to develop their teaching without threatening them.

In the afternoon, the group attempted a role-playing exercise. Some course members were given descriptions of the roles they were to play in a simulated staff meeting called to discuss the state of mathematics in a primary school; the Head of the mathematics department in a local secondary school had reportedly said that all he wanted was 'the basics'. After a few minutes' preparation, the group simulated the staff meeting, and the remaining course members watched the enactment. The instructions given to the participants were as follows:

An experiment in monitoring

Head
You are the head of a medium-sized primary school, and have called a staff meeting to discuss whether you are doing the right thing in mathematics.

There are reasons for this staff meeting — you may disclose them to your staff or not:

(a) At a recent meeting with the Head of Maths at the secondary school you feed into, he stated that all he wanted was proficiency in the four rules, fractions, decimals and percentages. Rumour also has it that the secondary school has been saying that you dabble too much in mathematics, and that you should concentrate on the tables and the four rules, and leave the frills to the secondary school.

(b) Your scheme is based on Fletcher, but you also use Hesse's *More Practice*.

(c) Your children have a test when they get to the secondary school; it consists almost entirely of the four rules.

(d) The recent HMI Report suggested that when children had a broad experience of mathematics, their mathematical attainment grew.

(e) The Adviser is urging you to get calculators and use them in the school.

Your mathematics co-ordinator knows that he is going to take action following this meeting. You fully support him and have delegated the job to him. Personally, you are not too sure about Fletcher maths — you remember the good old days of learning tables by rote — but you are prepared to accept the staff's opinion. You have 30 minutes to attempt to get some agreement on whether you are doing the right thing or not.

Mathematics co-ordinator
The Head fully supports you in the use of Fletcher and Hesse, even though you feel that he would like more of the formal work.

You would like to introduce even more 'Fletcher'-type work, but you know that this meeting has been called to consider whether the school is doing the right thing in mathematics.

Your task, following this meeting, will be to put its decisions into effect. You are aware that much material is used in the infants classes, but not so much in the juniors — this worries you, and you personally believe in children using material — especially games.

You also have the task of helping colleagues, especially the probationer.

Probationer
You have recently joined the school; your teaching practice was in a school where they drilled the children in tables and four rules. Your point of view is that the children are really hampered unless they know their tables. You see your job as making them learn them — by rote if necessary — so that they can do other mathematics.

You do not see the point of letting the children use the arrow symbol used by Fletcher. In fact, you do not see why they could not just use Hesse.

An experiment in monitoring

Make your point of view felt. You were a mature student, and your husband has recently been appointed Head of Maths at the secondary school. (You don't need to mention this unless you feel it will strengthen your argument.)

Mathematics Adviser
The Head has asked you to come to the staff meeting because he is concerned about whether the school is doing the right thing in mathematics. You realise that different views are held within the school, which uses a mixture of Fletcher and Hesse. You know that the children have to take a test on the four rules as they enter the secondary school. You would like this to change, and wish to encourage the two schools to get together more.

Personally, you are extremely enthusiastic about introducing calculators into primary schools, and have a project in another part of the LEA which is showing that calculators do allow children to develop number skills, and increase their confidence.

Infant teacher
You work in a team situation with a colleague. You both use a lot of material, and you don't go as far as HTU except for very able children. You know that there is very little use of equipment in the lower juniors; this concerns you, and you would like to see much more use being made of equipment of all sorts, especially base ten material. You believe that children need this in order to get pictures in their minds of the numbers they are using.

Make this point of view felt, and try to convince your junior colleagues that children still need a lot of equipment and materials for concept formation.

Lower junior teacher
You work in a team with a colleague. You both believe in weaning children away from material, for at least two reasons:

> the children themselves think it's kid's stuff; you want them to get abstract ideas, and the best way is to do away with materials and commit ideas to memory.

You spend a lot of time getting the children to learn their tables, and they know them when they leave you — or that is your belief.

Upper junior teacher
You use Fletcher a great deal and the children really enjoy their maths. Perhaps some of them don't know their tables, but you don't see the point of destroying good attitudes at this stage.

You are aware of some parental concern that this school does not prepare the children for the test that the secondary school sets on entry. It doesn't really bother you; you believe that the secondary school fails too, and that children at 14 or 16 still don't know their tables. They also do algebra and stuff that is likely to be no use at all to most of the children.

You also take an adult education class for literacy and numeracy, and there you meet people who have an utter disregard for education and teachers — they had a bad experience at the hands of teachers who made

An experiment in monitoring

them do unpleasant things, and you are slowly but surely putting things right for them. You believe in finding out what children and adults are good at and reinforcing it.

This activity was designed to encourage discussion about the range of current views about primary mathematics. However, it was generally thought to be unsuccessful because it was not taken very seriously. As one co-ordinator wrote in a report to her colleagues:

In the afternoon session we did a role play of a staff meeting on mathematics (which turned out to be hilarious).

Another teacher wrote:

Luckily quite some way from my own experience.

Perhaps role play is more useful when there is no audience, and the participants themselves can concentrate on understanding the roles, rather than playing to the gallery. This was probably a new experience for the teachers, too, and it might be more valuable on a second attempt. However, one co-ordinator did find it useful — she thought it aided awareness of liaison problems with other schools.

During the final session of the day, the range of apparatus needed in a primary school was discussed.

3.4 The second day

During the second day of the course, one session was devoted to encouraging the teachers to help children to calculate mentally as well as on paper, and another concentrated on games and puzzles. The teachers also looked at and discussed material they had been asked to bring to illustrate their own school's schemes of work and methods of record-keeping. This provoked the most challenging discussion so far, although most of the discussion still seemed to be on the level of 'what' was being done, rather than 'why' it was being done in that way. Problems were aired, such as:

I have a scheme, books ... materials ... but our HMI still isn't sure we have progression ... There is a need for a wide range of books to get varied experiences, and yet there still may not be progression ... what is impeding this development? ...

You can't turn non-mathematical teachers into good ones.

Yes, you can ...

No, they are threatened by schemes and records ...

... there is a fine line between helping, supporting, and giving a barrage of things ...

The whole issue of the threat presented by change came into the open,

An experiment in monitoring

and the teachers discussed how to affect people in such a way that they would work differently with children:

> A personal example ... of mathematics teaching ... may lead to respect for the person who gave the example ...

> Match what is being done with the Guidelines ... put on a display ... organise a year project ... to show good work, and to 'shame' ... organise a joint display for the parents, to show progression ...

> The development of personal relationships is essential ... broaden the experience of the person with children ... suggest omitting formal mathematics for a term ... try team teaching ... try specialist teaching — but many people lose out if you do this ...

This session, and the earlier one devoted to equipment, encouraged some of the participants to look again after the course at their own school's schemes and apparatus; however, the session itself did not strike questionnaire respondents as being either particularly useful or as useless. One teacher stressed that:

> The discussion can bring out aspects of a situation which, although important, are not immediately obvious.

3.5 The final day

One session on the third day of the course was devoted to assessment. The teachers broke into small groups to discuss the following questions:

1. How does the Head/class teacher know:
   whether the children are learning what they have been taught?
   whether the children are being given appropriate work?
   whether the standards in general are right in the particular school's context?

2. What objective information do you have on individual children or on standards in general to form the basis of discussion with:
   parents?
   receiving schools?
   children themselves?

3. When tests (formal or informal) are used in a school, which of these purposes do they serve, and what happens as a result:
   ordering children
   assessing the performance of individuals
   diagnosing weaknesses to see what work children need,
   comparing one group with another,
   comparing this year with another year?

Members of the group that the researcher joined found it very difficult to be analytic about these aspects of assessment; they polarised between supporters of subjective and objective approaches to assessment. General discussion focused on the difficulties of diagnostic testing and of helping teachers who cannot diagnose.

During the second session, two heads of mathematics departments in local secondary schools spoke to the course about liaison between primary and secondary schools. No doubt the role-playing activity on the first day had prepared the teachers to expect very different people from the friendly, concerned and supportive heads of department who had been invited. When they recovered from the shock, the teachers were able to raise some of their worries with the secondary teachers — about repetition of work in the first year of the secondary school, and about the types of record that would really help the secondary school.

Finally, the teachers made plans for future action they would undertake to follow up the course. They were asked to write down two things they would do as a result of having been on the course. These were gathered together on the overhead projector, and they included:

In-service in the school
Staff discussion — practical work, not computation only
    — move away from too early recording
More mental work and discussion
Survey mathematics equipment in the school
    — pool and maximise its use
Use calculators more
Encourage staff to swop classrooms with me, use their strengths
Take groups from other classes with difficulties, and do practical work and games
Devise a diagnostic test for children of 7+
Find out what maths is going on in the school and devise a scheme
Liaison with the junior school — swop teaching once a week
Discussion leading to scheme leading to record keeping
Think about my personal role as a mathematics teacher
Liaison with our secondary school — at least go there
Revise list of staffroom resources and make staff aware of them
Take a very hard look at our mathematics teaching

Some of these planned tasks were ones which the co-ordinator could easily carry out as an individual working alone; others involved trying to persuade colleagues to develop their teaching. How successful were the teachers in keeping to their resolutions, and did the course have any effect in their schools? The follow-up questionnaire to co-ordinators and their Heads tried to find this out.

3.6 After the course

The seven teachers and their Heads who replied to the questionnaire, either by post or orally, reported on the effect of the course on them and their schools in the term since the course had finished. The most immediate task for several of them had been to make a list of the mathematics equipment in the school, and they had begun on this. After a staff meeting to discuss the course, one co-ordinator in an infant school had set up a mathematics resource room in which to store and display practical mathematics equipment. This teacher had also organised time to enable her to teach some mathematics to all the top infants, and had spent time in other classrooms, working with groups of children. She had also begun to reorganise the school's scheme of work, outlining a sequence of work on balancing and weighing. Perhaps it was comparatively easy for her to do all this, as she was a very experienced co-ordinator who had long been responsible for the art in the school.

In a junior school, a staff meeting was held immediately after the course, and the next term the Adviser was invited to chair a staff seminar on how to make mathematics more practical. The Head also noted that some specific new activities had been introduced after the co-ordinator had seen them on the course.

An infant co-ordinator explained how she was re-thinking her own mathematics teaching 'hopefully to be more centred on a child's real needs and interests', and how her philosophy was developing; she continued to ask herself 'Why do I teach a particular thing, and could I teach it in a more relevant way for the children?' She had continued a programme of school-based INSET in which she led staff discussions on curriculum development, equipment, methods and the approach to mathematics. Neither she nor her Head reported any change in the co-ordinator's activities since the course, but the Head added 'This does not mean that I am dissatisfied with what is being done!'.

One co-ordinator, who had thought the session on liaison 'super', had visited the secondary school to which her children transferred. She had also found the list of possible tasks for the co-ordinator very useful to reflect upon. Her Head described her as lacking in confidence, but she had found the course very reassuring, and felt that she was growing personally through having to do the co-ordinator's job. She would have liked regular co-ordinators' meetings to enable her to meet people in a similar position, and felt that after-school meetings were not adequate; a day was necessary in order to give time both for input from the Adviser and for discussion.

One teacher who went on the course was the Head of a small school, rather than a co-ordinator; his school had no co-ordinator, and he had not yet grasped the need for co-ordination of the mathematics teaching. When he applied, he did not realise that the course was for co-ordinators, but went on the course to make sure that his school was 'doing maths right'. Since the course he had introduced more mental work, and had reorganised his class, so that he could teach groups and use apparatus. He had realised that his school did not have enough apparatus, but he still did not know what the other teachers in his school were doing in mathematics. However, he had now begun to plan a school scheme.

The last teacher to report on his experiences had only been out of college for two years, and had only been formally appointed co-ordinator in his school after the course had finished. He had tried to tell his colleagues about the course at a staff meeting, but it had been difficult — there were few links between the infant and junior parts of the school, and his older colleagues gave the impression that they knew it all already. He continued to feel unprepared for his new role; he needed to establish a different relation with the staff, and said he needed much more discussion with the Head and Deputy, in order to clarify the role. He felt that he had been given the job and left to get on with it. Personally, however, the course had made a great impression on him, and he frequently expressed his enthusiasm and personal gain to the researcher. His summing up was that the course had changed him, but not yet the school.

3.7 Discussion

The primary purpose of this course was not to change the way the teachers carried out their own mathematics teaching — it was assumed that the people who came on the course were already teaching mathematics in an enlightened way, and that their chief needs were to develop their conceptions of their role and to develop their inter-personal skills in dealing with colleagues. Most of the respondents claimed that the course had not changed their view of the co-ordinator's role, or the range of activities they were undertaking as co-ordinators. However, they came away with a good deal of information which might strengthen their work. They knew that the LEA and the Adviser supported their efforts to move their schools in the direction of more practical mathematics, more mental mathematics, more discussion and more use of the environment. They knew that the LEA thought that a school scheme of work and progression were important. They had thought about some of the difficulties in dealing with their colleagues. In consequence of these experiences, they all undertook some new activities.

How far their inter-personal skills developed as a consequence of the course it is difficult to tell. This is not something that can be assessed by means of a questionnaire, nor do skills in dealing with colleagues develop overnight. It is not clear whether the activities of the course were fruitful ones for developing inter-personal skills — or indeed if any course activities can develop these qualities. Certainly, sharing problems and finding that others had similar problems had a beneficial effect on the teachers' morale.

An experiment in monitoring

In general, knowledge helps to bring conviction and confidence; in a curriculum area where so many primary teachers lack confidence, the confidence that the LEA supports the co-ordinator's efforts and the knowledge that official documents such as *Mathematics 5–11* suggest lines for development could be very important to a co-ordinator, who might otherwise suffer from a feeling of isolation and who might be worried by the indifference or hostility of colleagues. Moreover, not all co-ordinators are as enlightened in their own teaching as are the best, and not all schools have gone as far as the best in their mathematical development. Contact with those teachers who have been able to go further, or who have gone in different directions, can widen the co-ordinator's horizons and encourage the development of the school's work.

CHAPTER 4

Case studies — a group of two-day and three-day courses

4.1 Introduction to the case studies

After the first experiment with a questionnaire, interviews with the teachers were conducted on a less formal basis. Most of the interviews were taped and transcribed for further study; in a few interviews, and during the course sessions, the researcher kept detailed notes. In studying the transcripts and notes, an overwhelming impression came through of the vigour of the teachers' expression, and the freedom with which they have been willing to talk about their work, their impressions of courses, and their INSET needs. As a result, in writing up the case studies, a style has been used of letting the teachers speak for themselves, rather than paraphrasing what was said in an interview; the case studies consist largely of quotations from what the teachers said during interviews.

Discussions with the interviewer took place in confidence, and so it is important to preserve anonymity, and as far as possible to prevent identification of particular courses or providers. Consequently, all names and some factual details have been changed. In the quotations from the interviews, what was said has not been changed at all. In quoting from conversations, an attempt has been made to preserve the rhythms of speech, rather than the conventions of the written word. Occasionally, for the sake of clarity, a few words have been omitted or added, but the intention has always been to preserve the sense and emphasis of what was said.

It has been a considerable problem to know how best to select from a mass of material, while retaining the flavour of in-service work and of the teachers' reactions to it. It is of crucial importance to present a balanced picture, and the picture presented is inevitably impressionistic. When different teachers have expressed contrasting views to the researcher, the range of opinion has always been included. However, although the researcher usually was able to interview the majority of the teachers who attended a course, in no case was she able to talk to all of them, and other teachers might have expressed different views. In a few cases, teachers or their Heads were unwilling to be interviewed. This may have been due to pressure of work, or perhaps it was inconvenient for the school to receive a visit at that time. On the other hand, a teacher's unwillingness to talk to the researcher might have stemmed from a wish not to be openly critical of a particular course, or to speak about an unpleasant experience, or one that she felt had been a waste of time. In all cases, permission to undertake a case study was given by course organisers, but they did not know what would be found, and course organisers were of course not given any feedback about individuals on their courses by the researcher. However, some teachers might not have grasped that the research was entirely independent of the course organiser, who was in some cases their employer, although letters to schools requesting interviews made that very clear. This might have contributed to the reluctance of a few teachers to be interviewed, or to teachers being less critically frank than they might have been. However, in general the teachers seemed to be very frank and open about the difficulties as well as the successes of INSET.

In each course, some features seemed to dominate the teachers' comments; often, these features also impressed themselves on the researcher as she took part in the course. Consequently, each case study tries to bring out those features which seemed to be special about that particular course. Some of the material that was recorded on tape was not directly relevant to this study; some discussions were relevant to the role of the mathematics co-ordinator, to the relationship between the mathematics co-ordinator and the head teacher, and to general problems of organising mathematics teaching in primary schools. However, the project did not set out to study any of these topics, and although much has been learnt about them, the gathering of information about them has inevitably been unsystematic, vitally important though they are. The project's concentration has been on the role of *courses* in in-service education in primary mathematics. Thus, the case studies illustrate some of the interactions between courses and teachers, and some of the aspirations and problems of course providers.

The first group of three case studies is of a group of courses with rather similar structures: they were all residential two-day or three-day courses at local residential centres run by LEAs. The format of the two-day or three-day residential course is popular with providers. It is widely thought that the intensive nature of the experience is very valuable in promoting enthusiasm and raising morale; moreover, it is possible to do much more on a residential course in two days than would be possible if the teachers travelled each day. The experience is also a social one, and assumes something of the air of a mini-holiday given to the course members by the LEA, as well as providing an intensive period of concentrated work on one aspect of teaching. Consequently, many LEAs maintain residential centres for their teachers, although some have been closed in the recent educational cuts.

4.2 A weekend residential course for infant teachers

4.2.1 Course organisation

The course took place at a LEA residential centre, and lasted from Friday evening to Sunday lunchtime, one weekend in mid-July. It was intended for teachers of the 5–7 age-range; however, a few of the 40 teachers who attended taught older children, but included infant mathematics among their responsibilities. Most of the course members were resident at the Centre. The course was heavily oversubscribed, and so one place only was allocated to applicants from each school, although one teacher described how she wangled a second place:

> We both sent our names in, and I just got in on a cancellation, because there was only one place per school . . . I rang up and asked County Hall . . . I said if they did get a cancellation I could just drop everything and come. . . . And it was 2 o'clock on the Friday, so it really was a last-minute thing . . .

The LEA Mathematics Adviser had invited Margaret, a lecturer from a college in another area, to direct the course. On Friday evening she spoke to the course members about making use of the environment and the children's experience in infant mathematics. She described with many apt illustrations how classroom experiences could be structured to encourage discovery, and how contexts could be used to suggest problems. She contrasted two infant teachers, both of whom were committed to practical mathematics, but in very different ways. One used Unifix as an aid in working through sum cards. The other, whose class had done a lot of work on potatoes, asked the children to make potato prints at random and then in sets; discussion produced the equivalence of '2 lots of 6' and '6 lots of 2'; one child noticed that hers was 'the same both ways'; this type of activity, Margaret said, was indeed 'the music of mathematics'.

Part of the Saturday was spent in applying the idea of finding mathematics in the environment practically in the grounds of the Centre; following this, displays of the teachers' work were mounted. Another theme was the analysis of progression through a mathematical topic. Length and weight were the topics studied, and the

Two-day and three-day courses

teachers worked in small groups to order the progression of development through their topic. On Saturday evening, Margaret had invited an infant teacher whom she knew to discuss and display work that she had done in her classroom. On Sunday morning, there was a display of mathematical games suitable for infants, and the teachers worked in pairs, playing the games and discussing them. The course dispersed at lunch-time, with plans for local follow-up sessions in various centres in the LEA.

The researcher attended the course as a participant observer, and in the autumn she visited the schools of seventeen of the course members and discussed the course with the participants and their Heads. Thus, slightly fewer than half the course members were interviewed. Opportunity was also taken to discuss in-service work more generally with the teachers, as well as some of the problems of infant mathematics.

4.2.2 Reasons for attendance

The course attracted a wide range of teachers — young and old, habitual course-goers and those who rarely go to any kind of course, confident teachers with responsibility for mathematics, teachers who badly needed a boost to their confidence, idealists and cynics. Some of them told the researcher about their reasons for going on the course:

> Since I took on the Scale 2 post I have been wanting to go on a maths course where I went into a little bit more depth than you can on evening courses, just after school. I didn't know what other schools were doing, or other people with posts of responsibility, and I just felt it would be a change to meet other people and see what they were doing.

> The Head asked if I would like to go because of my responsibility for infants, and because maths is not really my area ... and it really was EXCELLENT. I'm not so much unsure about the maths, but about how to teach it — and making sure they can do it.

> I wanted ideas and felt I lacked inspiration ... I did get what I wanted — ideas and ways of using things, especially the games on Sunday morning. Other things revised what I had in a good college course, and reinforced what I am doing. I need a course about once a term to keep me going.

> I've never felt maths was my strong point; in the classroom I've always found ... not so much limitations in myself, but a need for back-up. I was pleasantly surprised with what I found there, and in fact it extended me a terrific amount.

> I went on the course purely because I thought I needed a bit of wakening up; I haven't been on a maths course since I lived in the north east — and that was twelve years ago.

> When I go on a course I want to be made to feel that I want to look again at what I am doing — justification, aims, objectives — we perhaps don't

> think enough about why we are doing what we are doing. When I started teaching, I didn't have anyone there to tell me if I was going along the right lines — if I was doing all right. You know, reassurance means a lot. You go on courses partly because you begin to find out your own black spots and inadequate areas, and partly because you want to find out how other people approach things.
>
> As I pointed out, I am rusty ... the first residential maths course for a long time ... yes, I think it's my first residential maths course. It keeps you aware of what is coming out — if people are producing anything new ... and seeing — if you like — what the party line is at the moment ... I don't mean necessarily toeing the party line ...

Some had been anxious and apprehensive at going on a mathematics course, and there was even one teacher who, having been successful in obtaining a place, could not bring herself to go, as her Head told the researcher:

> Two from the school applied — in fact it was the other infant teacher who noticed the course first. However, only one place was offered and the other teacher was not sufficiently secure to go on this course on her own, and so Mrs Perkins took up the offer. They are going to run the course again, but I'm afraid Mrs James will still feel too insecure to face other people right across a weekend, although she is prepared to go to evening courses, and even though Mrs Perkins enjoyed it. (Head)

Another teacher told the researcher about her worries before the course:

> I don't know very much about maths, and I was worried when I went that I was going to be in with a lot of people who knew much more than I did. But I didn't feel like that at all — I was made to feel at ease.

One course member did, indeed, have an unfortunate experience of being mismatched with her partner, but this did not, in the end, prevent her from valuing the course:

> On the Sunday morning, for the games ... I was unfortunately going round with a maths expert or someone who had done a degree or diploma or something — and she didn't give me time to work anything out. And I found this very annoying, because I have to see things — I'm not very quick, I need to work through it. But I was fired with enthusiasm by the course — and I did feel reassured, actually, and I could see there was an awful lot to be done.

4.2.3 Looking back on the course

All the teachers interviewed judged the course to have been a success. Much of the success was attributed to Margaret's personality:

> I have never met anybody like that — not for years. She was an amazing woman. She was a very strong personality but she was so approachable, and such a good lecturer but not in a threatening sense.

Two-day and three-day courses

> I had never met her before; we all liked her very much. I thought she was really full of enthusiasm, and I felt immediately full of enthusiasm myelf. I thought she was a very good speaker. On the Friday night when we got there, we could have felt very tired after coming from school and getting to the course, but she immediately got me very keen ... even if in some ways it was because I violently disagreed ... but at least all of a sudden you seemed to be into it, and I was thinking about it. I felt straight away that you could disagree and it wouldn't matter, and that you could go and talk to her about it.
>
> She was such a good speaker as well — you can get someone who knows their subject inside out, but with a large group of people cannot communicate ... and she was so good with her organisation, and she didn't threaten you — she was just right really. I have been on courses where it's way above my head, and although I don't mind, I think I am really wasting my time.
>
> The people who were running it were exceptional; it was excellently planned. And they were all nice and friendly and relaxed ... and I felt we were all there for the same aim — to get as much out of it as possible .... and you put in to get out, don't you?

The visiting teacher was also much appreciated, and the quality of her work provided an example, as well as some much-needed reassurance. The need for reassurance came strongly through many of the interviews, and only a practising teacher could provide reassurance for those who were doubtful about the practicality of the approach advocated on the course.

> I suppose the general atmosphere of the course is what sticks out most in my mind ... but I think that woman coming in and showing us all the stuff she had done in her classroom was the main thing. To see other people's work is very stimulating, and also it makes you feel 'Oh, well, perhaps I am not so bad after all, and I had that idea too', or 'That's a good one — why don't I play on that?'
>
> I think the course was not so much a refresher — I have had a lot of ideas for doing things like that, and indeed I have done a lot of those ideas. When I worked in the north we did a lot of practical work, but I think you always doubt yourself after a while, as to whether it is really worth while. And when you try to get it over to other people, they have doubts — and they say 'Are you sure these ideas actually work?' and 'You have children running around all the time, and you can't really see what's going on'. So as far as I was concerned, I went there and I saw a teacher doing these things, and it made me really feel 'Yes, it does work after all', and I came back revitalised.
>
> Margaret was saying all the time 'You don't have to be a maths expert', and the teacher she brought in on Saturday — I could see a bit of myself in her — and she was saying she hadn't been a maths expert, and still isn't, but she is able to think of all these things to do.
>
> I was very impressed by that lady who came with all her work from the school. I found it marvellous that she could get all that work out of her infants ... I thought that was impressive.

Two-day and three-day courses

> It was reassurance, and to make sure we weren't doing anything wrong ... but then of course after that it was the end of the term, straight away ... but you keep something from it, you keep it in mind and you use it where you can ... it did teach you to look more for maths in places where perhaps you wouldn't have looked for it.

The course was intensive and, because for most of the teachers it was residential, there was no let-up even outside the course sessions. This was by design, but the teachers' stamina varied; although some lasted well, there were others who were beginning to fade before the end.

> I did feel the whole weekend that we were actually doing something. I sometimes feel that when you go on courses after school, you are tired after a day at school ... and if there is a speaker, you can concentrate for twenty minutes or so, and then your concentration lapses, and often at the end of the evening you think 'Well, I didn't really get much out of that' ... But with the weekend, I felt stretched ... I thought it was quite hard work. I feel that courses after school don't really stretch you enough, and I felt that we achieved something ... we were working all the time, so that I didn't feel that I had wasted any of my weekend, and yet all of it was quite enjoyable ... it was fruitful from that point of view.

> I often go home and tell my wife after a course ... well, either ... the company was marvellous and the content was awful, or vice versa ... rarely do the two seem to go together ... but that one I would have said was enjoyable in all respects ... and valuable in all respects.

> It was tiring ... it was too much ... I felt ill on Monday ... dizzy ... It was not so much the course, but that one never stopped ... meal times ... at least I had a room of my own and did not go on discussing at night ... one needs a break between the end of the course and school ... Monday morning off!

> I thought it was bit too intensive ... I was utterly and completely drained. I would have appreciated a little more time off, because in the end I felt I couldn't cope with anything more — but I thought everything was most worth while. I think my mind could take in so much, and I had reached a point where I had to reject something, so I rejected those games ...

4.2.4 *Development in teaching*

Some of the teachers who were interviewed felt that their teaching had developed considerably as a result of the course. By the time the researcher came to visit them, they had had a term or so to think and to experiment in their classrooms. The most noticeable developments were a greater use of the environment, and a more relaxed attitude towards mathematics.

> I have been more practical, had more discussion, and gone slower since the course. I have done more from the environment ... but it's still early

stages ... I'm re-learning my own maths, and I still need convincing. I would have thought what we did this morning a waste two years ago, because they had done no writing. ... but I feel guilty and there are so many pressures. I'm still frightened that I might be going too slowly ... it's all pressures within myself.

However, achieving the right balance between environmental work and other aspects of mathematics remained a source of anxiety to the teachers; Margaret's enthusiastic presentation of environmental work had caused some of the teachers to misunderstand her advocacy of it:

We have got the Nuffield and the SMP, and we don't tend to use environmentals ... but I think someone who is very capable and knows where they are going could find it easy to draw the mathematical value out of everyday experiences ... at first I thought she was suggesting that most maths could arise out of the things that were happening, and that it was better if it did ... but talking to her afterwards, and saying that we felt how much we did need something like a scheme to rely on ... then I did realise that she was showing us one aspect, and telling us a lot about that, and I understood that then.

A Head who had not been on the course described how she was able to help her teacher to see that work she had previously done with formal apparatus could more appropriately and vividly be done with the everyday materials in the room.

Mrs Perkins is saying that she realised through the course the value of the environment ... and that came through for a lot of people ... it also came through that some of them thought Margaret was saying you should teach entirely through the environment, and she wasn't ... But Mrs Perkins talked to me this morning about something she had done with the toys ... and she said 'I got the children to look at them, and then we talked about them, and then we put them into sets', and actually she was talking about a whole philosophy ... and she said 'At first they could not even describe them — couldn't even talk about them', and then I said 'The work we do on attributes, when we actually structure it, we are putting too much emphasis ... but you are bringing out the attributes with those toys'.

(Head)

Another teacher was seeing new opportunities, both in topic work and in everyday classroom incidents:

One can do topic work of some sort, and touch on the maths, but I think the course showed me that you can do more than touch on it — you can actually get more maths out of it than I ever thought was possible, and that was very helpful ... For instance, Catherine came in with a scarf that went six times round her head, and one little girl had one that hardly went round at all, and so we had both scarves out in the classroom, and we tried to decide whether one was TOO long ... one was TOO short ... I must admit that at one time, I would have said 'What a lovely long scarf' and left it at that. But now I pull out anything I can.

Two-day and three-day courses

For one teacher, the course came at just the right time; having got her language work organised, she was ready to look at mathematics, and to move on to experiment (cautiously) with a new teaching style.

> I needed something new to latch on to ... and it was just a coincidence really ... there were all sorts of coincidences ... I was able to get on the course ... and she was an extremely good speaker ... and it came at a time when, as I said, I needed something new to start thinking about.
>
> I was happy with how the reading was going, but I wasn't with the maths ... and then, having been on that course, and seeing all sorts of other possibilities ... in a way it gave me confidence to try doing things. I think I thought at first that, if I did nice things with them, they wouldn't be learning maths at the same time ... then I realised that they would be learning, but I hadn't realised that I would have to keep a check on what was going on ... having been on the course gave me confidence, and then I thought I couldn't possibly have them all rushing round measuring at the same time, so I have got to do something else ... so I divided them up into three groups, and there were three tables where there were different activities, and they would circulate round ... and I have got them trained now ... and they come up and say 'I have done this, where do I go now?' ... If we worked a truly integrated day the children would choose ... but I don't think we could do that ... so I have got to be half and half ...

She was also more able to make use of opportunities as they occurred, and she gained confidence from the successful way her new method was working, and from the interest and approval of her Head. She found herself doing a lot of work on 'pairs':

> I hadn't planned it — it just came spontaneously from a TV programme ... we went out and found pairs of leaves, and looked for pairs of things; we looked at ourselves, and what pairs of things we had ... what has been so valuable is that the brighter children are being quite stretched, and I am sending them off and they are counting things in twos ... and of course the other thing that has been so useful too is that they realise that you can't always make pairs ... In the Hall, we found partners, and of course we found that sometimes there is a child left out ... then with their partners, they took it in turns — one would make a shape and the other one would try to make the corresponding shape ... symmetrical ... you don't appreciate sometimes with little children — how able they are ... And we are going to do it again in the Hall for the older children to watch.

She then looked back at her previous style of mathematics teaching, and contrasted the work she was doing now with it. Because she could see greater success coming from the less formal work, she would be able to develop from there, although she still had some doubts.

> Last year ... they just worked through the maths sheets, and if there was a problem I would just stay on that level for a while, and give the child activities round that. I have always done a lot of sorting, but what I called the maths activity of the day was cut and dried and finished, and it isn't any more.

Two-day and three-day courses

> What seems to be happening now is ... I still do the Scottish maths twice a week, so that I can keep a check on how they are getting on ... and they are getting on fantastically ... I know that you can't compare different groups of children ... last year's group were different ... but these have got on a lot more than my children last year ...
>
> I would like to see how it works with children I am less confident in, and with children who might be more difficult, and with a larger class — this term I have't got that many children — only twenty-one ... but whether I could keep this up for a whole term ... I'm obviously doing formal work sometimes ... but whether I could keep up the sort of mess of it all with a larger number I don't know.
>
> I think probably the games helped me as well ... drawing things from the environment has meant that I have changed my sort of maths teaching ... and I have been able to make the materials for some of the games ... and I now feel the children are more meaningfully occupied most of the time ... purposeful games ... and so I would say both of those have helped me — sort of on an equal level.

Some of the other teachers, however, although they found the games interesting, did not follow them up in the same way as they did the environmental work. Perhaps this was because the games needed more input from the teacher, in making them and teaching them to the children, while some of the environmental work was in the form of a spontaneous response to the opportunities that presented themselves in daily classroom life. This was certainly not the case, however, for the teacher who had re-organised her classroom so that three tables of activities were always available. For her, games gave another very useful source of available and purposeful activities. A selection of comments on the games from other teachers follows:

> I was very interested in those logic block things ... I was very interested in the games ... and how difficult they were ... I was astounded by that ... there was a lot of thinking involved. I did a bit of it after the course, and although it wasn't an on-going thing — it tended to die off after a while — more from my fault than theirs ... I could see that there is a lot of potential in it.
>
> Yes, I've done games ... I think some of those games ... I wondered about ... now what do you call them ... those logiblocs ... I tried that with mine and they catch on ... they can do 'one difference' very quickly ... and that's four-year-olds ... that's very good. I haven't made any of the other games ... but there's still time.
>
> I am with reception, so I am very limited at the moment. I would not want to follow up on those games, but I think if I were taking the older ones I would.

4.2.5 The progression of a topic

The remaining major activity of the course was a session of group work, in which the teachers took the topic of length or weight, and traced its development through the infant years. This evoked mixed feelings: for some teachers it was the high spot of the course, but for others it was a peripheral activity, which contributed little to their thinking.

> On Saturday morning we took a subject, didn't we ... I took length — in fact, we went through the progression of it. I thought that was very useful, actually, it got people talking a lot more, and thinking through a subject like length actually directed you to the way you are doing it ... in this school we always discuss where things should go ... If there was any follow-up to the course, I would really quite like to go through nearly all the topics and do the same thing that we did with length, which was really to think out where we are going with it ... I think something like that would be useful to do at a staff meeting, with the whole school ...

> I don't know about the other groups, but I felt that we never really got down to that — we sort of got sidetracked into putting things on bits of paper, and didn't get on to the talking.

> The ordering discussion ... that was the one where I most had to think ... newly ... the others fitted into what I would already see as my general attitude, anyway.

> The planning exercise was really most useful ... it really made you think ... I did length ... at first it seemed easy ... and then one realised ... It would be a good exercise for a staff meeting — I mentioned it to the Head — even when you think you have agreed you realise that someone else means something different — and that could happen with the Nuffield, too.

For one teacher, now a successful and enthusiastic mathematics co-ordinator, the activity evoked unhappy memories of an earlier style of mathematics teaching, before she discovered environmental and practical work. She related to the researcher how her conversion to being a 'maths enthusiast' had come about.

> That planning we did — I don't think that was entirely useless. I think that in itself it made you think of just how complex things are ... the more you investigate maths ... the more you work on maths ... the more you realise you don't know, and the deeper you go the more holes you find ... perhaps that's why some teachers are afraid of it ... maths can get very complex — I used to be scared of it myself — when I first started teaching I hated it. ... If you think in terms of colours, language and art work were the very bright colours, maths was always grey and black.

> I was working in an open-plan school, and we did a lot of topic work, and having developed the 'research' side of things we started to put other areas of the curriculum into it — it wasn't just language and art based. We did a term's topic on birds, and the amount of maths we brought into that was unbelievable — that was the first really successful project that maths worked into so well and easily ... and we found to our surprise that it really did work.

Two-day and three-day courses

4.2.6 Reporting back

In some schools, it is the custom for a course-goer to report back to colleagues after the course, passing on ideas and knowledge learned on the course, and so providing some school-based INSET for colleagues. This was a very difficult course on which to report back — so much had depended on the atmosphere and the enthusiasm generated by the lecturer, and by the course members among themselves.

> Her reaction was that she felt that for other people to benefit they would need to do the course, rather than to share from her ... I don't really know what she meant by that ... not having been on the course myself.
> (Head)

> Well, so much was said on the course ... it would be difficult to transmit it. I think I gave them the feeling ... but it's difficult, isn't it ... it's a very personal thing ... we spent a lot of time on the course doing it ... so you can't just say 'We found this and this and this.'
> (Teacher)

> I have just asked her how she enjoyed it ... She just said she found it interesting — frustrating in some ways. But I think she has a lot to learn about infants.
> (Head)

> I have talked to some of the others individually ... I've told Mr. Paul [the mathematics co-ordinator] about it ... senior members of staff do give reports ... but there isn't much feedback really.
> (Teacher)

The pair of teachers who had come from the same school told the researcher about the advantage of going as a pair. However, they had failed to share their experiences with their colleagues — perhaps the incentive was less when they could discuss it with one another, rather than with sceptical and unconvinced colleagues.

> There is an advantage in going in a pair from the same school, because when you get back you can discuss things in the atmosphere of your own school, and how it will apply to you ... it is very difficult to be one on your own ... if you come back, and you are the only one, and you are very enthusiastic about something, you do tend to get cried down by the others ... I think I came back rather depressed, because while we were on the course, it seemed fine — marvellous — but when you try to apply it to your own situation you have got to find your own solutions ... and for various internal reasons we have got to have a very structured timetable ... we are running a remedial group, and I feel that my maths has to take place between there and there ... and so there is never any chance of any follow-up.

The course had been held in mid-July:

> That's another thing, we came back and it wasn't a good time for reporting back to the others really, was it? We found that they were too tired

Two-day and three-day courses

and we were full of enthusiasm, and we thought... we'll wait and see what the response is, but I think they were just too tired to receive it.

4.2.7 Local follow-up

An innovation of this course was an attempt to follow it up at local Teachers' Centres within the LEA. The Teachers' Centres leaders had attended the course and, during the next term, one of them arranged a follow-up meeting at the Teachers' Centre. Other teachers, as well as those who had attended the original course, could also come. The leader chose the topic of mathematical games, because she thought that many of the games shown at the course were new to the teachers, and that they needed more experience of them. The Teachers' Centre leader asked two local teachers who had been on the course to prepare specially for the follow-up session. This was a good experience for them, and a useful introduction to contributing to INSET, for they found that they had something immediately practical to say. One of them discussed her experience:

> I found it very useful because the Teachers' Centre leader asked another teacher and me, both, to try out the games with our children before the meeting, so I can't look at the meeting in isolation ... I found it useful because I tried the games with my children and I got quite a lot from it. I thought that the meeting went well from my point of view. But I don't know whether someone who just came cold to those games and just played them there would have found them easy ... whether they would have found it as easy as I did ... having used the games with the children before. My immediate reaction was that it had gone very well, and that people seemed to talk ... I went around and talked to people ... about what I had done with the children ... even at the end of the meeting when we had finished, there were people standing around talking and saying could they borrow this, or asking questions. So I think that quite a lot of people had been generally interested.

Not all, however, found the follow-up as helpful as did those who were actively involved, and one teacher imputed motives to the course organisers that might have surprised them.

> I did go to the follow-up at the Teachers' Centre ... there's not a lot of practical work or games here so it was not a very useful follow-up ... I would have preferred something like the teacher who came to the course on the Saturday evening ... a sharing ... showing things we have done ... if someone is brave enough to do this ... I wanted ideas ... to see what someone else is doing ... we are exchanging things more here now ... getting more settled and feeling we want to get together.

> I presumed the follow-up was because the powers that be wanted to know what we were doing — I don't know if that was a correct impression. ...

4.2.8 Discussion

The success of this course was largely dependent on the atmosphere of enthusiasm and confidence which it conveyed to the participants.

53

Two-day and three-day courses

For one teacher, it came at exactly the right time to enable her to make a significant leap forward in her teaching. For others, it gave renewal and reassurance that they were already working along the right lines. For some, it provided an outlook which they might not yet be ready to translate into classroom practice, or which they thought — rightly or wrongly — was impracticable in their situations:

> I think I came back rather depressed ... when you try and apply it to your situation you have got to find your own solutions ... The practical side of it was so good ... yes, I enjoyed that. It was lovely to see all those pieces of work laid out there, but there would be nowhere where we actually could put anything like that ... when you come back to your own school, you think that would be lovely, and then you come up against people who don't particularly want things there ...

Ideally, for this teacher, follow-up would have consisted of detailed one-to-one discussions about applying the ideas, perhaps in a modified form, in her own school. Her difficulty was not so much lack of enthusiasm, or lack of willingness to attempt something new, but inability to see how she could apply the new ideas in an unsupportive situation. In some schools, the Head or the mathematics co-ordinator would have been very keen for the teacher to develop her teaching in a more practical and environmental direction, but this school's mathematics was based on a rather limited interpretation of the Fletcher scheme, and most of its recent effort had gone into moving over from the first edition to the second edition. Environmental mathematics was not high on its agenda. There was also a shortage of practical apparatus, as the teacher's colleague related:

> ... when we get to the weighing pages ... at the moment I have gathered up all the scales ... and I have got them all ... because we are rather short of equipment. I never realised before, and that really had brought it home to me.

The teacher who was so depressed at the possibility of finding her own solutions was the same teacher who had been so keen to go on the course that she had been willing to drop everything on Friday afternoon to go on a cancellation. She probably needed continued help, which could only have been provided by support from outside her own school — perhaps her only real solution would have been to move to another school. One teacher, alone, cannot do all that she would like to do; nor can she always fully adopt the ideas that INSET puts before her. However, there probably were things she could have done — some environmental mathematics presents itself daily to those whose eyes are open — other teachers who went on the course saw new mathematics in toys and scarves and pairs; Fletcher can be interpreted much more broadly, and constant pressure on the Head or the co-ordinator by determined infant teachers may even produce a new pair of scales. The course might have been more fruitful for these particular teachers, however, if LEA staffing levels were such that they could have received a follow-up visit from an Infant Adviser who was aware of their problems, or if an Advisory Teacher had been available to work alongside them and to give practical demonstration in their classrooms of ways of implementing the approach to mathematics advocated on the course.

4.3 A residential course for infant teachers

4.3.1 *Organisation of the course*

This course took place at a LEA residential centre, under the aegis of the Mathematics Adviser and the Advisory Teacher for mathematics. It ran from Monday evening until Thursday afternoon, one week in June, and most of the 38 members were resident. The course was advertised in the following terms:

> *Mathematics in the infant classroom*
>
> The programme will be similar to that of the successful course held eighteen months ago and will reflect the wide variety of learning situations at the infant stage. Particular attention will be given to the importance of language and the place of written work. Many of the sessions will have a practical basis.
>
> The course is intended for those infants' school teachers who would not necessarily include mathematics among their main enthusiasms.

The programme opened after dinner on Monday with a lecture by an HMI with special expertise in primary mathematics. The lecture emphasised the importance of teachers' decisions about priorities in infant mathematics: the right balance was needed between different types of activity. The importance of mathematical language was emphasised, as was the need for teachers to take their time and not to rush on to ideas whose foundations are not laid. Finally, the beauty and dynamic of mathematics were stressed.

On Tuesday morning the Advisory Teacher, Polly, started by describing mathematical work from 'ordinary things around us'; this was illustrated by work from infant classrooms. The rest of the day was devoted to a lengthy practical session, in which groups of teachers took environmental topics and explored the mathematics to be found in them. Each group was led by a practising infant teacher in the LEA. The researcher joined a group whose leader introduced the first topic.

> I've chosen dinner numbers — I've used it as a maths topic — because it bored me and I wanted to look for its potential. I'm trying to help the children to prepare for the new canteen system; next year they will have to choose...

Two-day and three-day courses

In the afternoon, the teachers worked on a prepared imaginary situation:

> I walked into the school hall one day in the summer holidays and there was a ladder, paints, colour charts and two painters — one was short and tubby, and the other was tall and thin.
>
> So, what could I get from this?

To add verisimilitude, the colour charts were available for the teachers to use. In this particular group, ideas were rather slow to flow; perhaps the teachers were not used to finding mathematics for themselves in a topic such as this.

After dinner, participants viewed the work of other groups; Polly described the exhibition as 'like a real life resource book.' On Wednesday, groups of teachers tackled mathematical topics such as subtraction and place value. They were asked to order the stages of progression in the topic and its development through the infant age-range. The work was based on unordered lists of ideas in the topic, set out on worksheets; for example, the subtraction list included:

> Can appreciate difference when comparing two sets.
>
> Can count back.

The leader explained to the group:

> Remember it is just a list with no order — the column at the side gives us a space to put a rough idea of the age to introduce it. Our brief is to sort out the progression.

On Thursday morning, two infant teachers described the style of working in mathematics in their classrooms and showed examples of childrens work. Another teacher gave a talk entitled 'What price a maths workbook?', which indicated that for her the price was too high:

> Children cannot keep still — they learn by doing — finding out about the world. If you have done it with towels, do you need to do it in books? If you are using your own ideas you have to think 'Now how can I consolidate?'. Teachers must THINK. There is very little questioning in a workbook.

The course ended with a discussion session based on a list of ten questions, followed by a few words from the Adviser:

> We wanted to give you a spirit, and I can see you have enjoyed it — the danger is that one goes away on a high and does not follow it up in school. The nice thing here is talking to colleagues; remember it can be done in other places too.

A great deal of planning and preparation had gone into this course; Polly had met with the teacher leaders for five evenings during a long period to help them to prepare to lead groups, and they had worked

Two-day and three-day courses

on the topics in their own classrooms. The worksheets were very carefully prepared, and the work that that groups did on the ordering of topics was typed and sent to them after the end of the course.

4.3.2 Monitoring

The researcher took part in the course as a non-resident course member, and during the next term she visited nine schools in the LEA whose teachers had attended this course. She discussed the course with these teachers and asked them what ideas from the course they had brought back into their schools. Teachers from these schools had also taken part in other INSET activities provided by the LEA, and these activities were also discussed.

4.3.3 Reasons for attendance

The tone of the advertisement, with its clear identification of the target audience, struck a chord with at least one teacher:

> Maths was always a dreadful subject when I was at school, and this was one of the reasons that I wanted to go on that course. The wording was 'people whose first enthusiasm is not maths' and I thought 'That's just me'. Since I've been back in teaching, now I've had my family, I've found that maths is much more interesting and I feel that I have learnt much more myself, and so I can communicate with the children.

However, there were some who came not because they were in the target audience of the course, but because they knew they would enjoy it. Perhaps the next teacher would have been better suited by a more advanced course, which would have taken her 'conversion' for granted, but her enjoyment no doubt contributed to the general success of the course:

> I came back and I talked about the course because I had enjoyed it very much and got a lot out of it, but really it was the kind of teaching that goes on here all the time.

> So I had a different attitude ... Some of them on the course were saying 'I think it is super and I would like to do this, but my Head wouldn't like it if I didn't do formal number work', and already they were envisaging problems ... But I knew that I could come back and do virtually anything that I liked and it would be acceptable, because people here are already converted to that way of teaching ...

4.3.4 Talk

Many teachers count the value of talking to others very high among their reasons for going on courses, and rate the communication with other teachers as one of the most useful aspects of a course. This

Two-day and three-day courses

course was no exception:

> We are only four staff in this school ... So we need to be careful not to get into a backwater, and we go on courses as we need to talk to people. It's hard to meet other schools — they don't often seem to want to ...

> Talking to the other teachers — it was marvellous because so many problems are the same. That was my only criticism of the course; I found there wasn't enough time to talk to people, there wasn't enough time in the small groups to discuss — we often got to a very interesting point and it was time to stop.

Ideas were exchanged, and teachers who lacked confidence in their mathematics teaching gained reassurance, even though the course might not have helped them very much to solve their perceived prolems.

> People were talking informally over dinner and in the evenings ... You are reassured ... You begin to feel 'Ah, yes, I am on the right lines'. You feel reassured, everybody else is going through the same grind as you — we've all got the same problems and the same worries — and you don't feel quite so much alone and isolated at the end of the day ...

> I think that for myself it is reassuring to hear that other people are doing the same sort of things, and you think 'Well, I am doing it right'.

> I think it affects how you cope with the problems. You get other people's ideas and you think 'Oh, great, I've got something similar and I'll try that', and as well it gives you more confidence because you realise the problems are so universal.

Perhaps, though, like-minded teachers talk together and so do not get jolted to look enough at different ways of working:

> We've got a central store of maths equipment, but in fact the classrooms are very well equipped in maths ... From what I heard on the course I ended up keeping quiet, after hearing from people who were sharing one set of scales between a whole school...

For some, the talk was not only for reassurance and the feeling that a trouble shared is a trouble halved. Some participants felt they needed the stimulus of talk at a higher level, and of being given an opportunity to think in a way that they may not have time to do in school, when the daily problems are so pressing.

> I think it was the business of actually thinking something through — whether it was multiplication, place value or whatever — really having to sit and think about it; that was the most valuable thing.

4.3.5 Involvement

For many of the participants, this course gave more than merely the opportunity for a good talk. It was almost universally rated a success.

Two-day and three-day courses

There were several reasons for this. First, the participants valued that fact that — unlike some courses — it was not only advertised as being concerned with infant work but it was full of the spirit of the early years of schooling.

> I liked it, I enjoyed it, it was very good. It was so down to infants — this is half the trouble with some of the courses I've been on. The people running some courses are junior-oriented; they say it applies to infants, you know, but it's not the same at all.

The teachers became very much involved in the tasks they did on the course. They may have seemed to the observer to be a little slow to produce ideas, but this did not prevent them from feeling they had achieved a good deal. The first project built up their confidence, and then came the difficult task of analysing a mathematical idea. One teacher described the day's work to the researcher in a headlong rush of enthusiasm:

> The second day I had a whale of a time — I really did — we were doing different projects and I was looking at nature and I thought 'Oh, I think I know what I'll do', and it was nice because I like doing that sort of thing with children. And as I was saying, the discussion was very good.
>
> In the second group I was with, we did very early addition and we got talking about the real concept of addition, and how much you need to do with the children before you get on to actual addition. We were thinking very practically — what a lot there is to do about addition. And the language — how confusing it can be if there are so many different ways of saying 'add'. And we came to the conclusion that you need to say practically all of them and there's no point in sticking to one. One person said 'I think I'll stick to one and then later I'll tell them the others', but the rest of us said 'No, you just keep saying it in all the different ways'.
>
> It was very nice because I think we were the only group who all worked together on one thing ... We got a long piece of paper and people were dotted along the paper doing the progression of addition, and we were shouting things down from the other end of the paper: 'Have you covered so and so, and where should it be' ... And I think that was the best part of the course.

Other teachers shared a similar experience:

> I think it worked quite well — I found it very mentally exhausting because we really had to think about what we were doing, and why we were doing it in that order, and when we should introduce certain things.
>
> With place value, we spent a long time sorting out which stages came when, because it was all jumbled; really putting it in order, and carrying it through, was so difficult because the stages overlapped so much.

They had also found it very useful to look at the work that other groups had done, and they felt at the end of the time that there was still more to be learnt from this:

> It was also valuable going round to see the groups after they had done all their work — looking at the displays and seeing work that you could do in school, and I got quite a lot of ideas from that . . .

> Yes, you spend so much time — each group in each room had spent so much time doing the topic — it would have been nice to have spent more time than was allocated, to talk to people in the other groups.

4.3.6 Teacher leaders

The teacher leaders had been carefully briefed for their work by Polly; this in itself must have been a considerable in-service exercise — they had several meetings and the teachers worked in their own classrooms on the topics to be discussed over quite a long period. The course members readily accepted that colleagues should act as group leaders, and they were very interested in the talks the teacher leaders gave about the mathematical activities in their classrooms.

> I was very pleased actually — having a teacher as group leader; Caroline wasn't . . . she didn't take the leadership too much . . . she did start us off but she delegated, she said 'Now this is what I have chosen to do, but I want to hear what you think about it — you are all teachers. What shall we do?' That was very nice . . .

> They were good teacher leaders, and it was good to have teachers doing this . . .

> The session where the teachers were talking about their own classrooms — I think that was the last day. There was the lady who actually admitted she had done something wrong! I was nearly going to shout 'Hurrah; somebody's actually said "I did this, and as I was doing it I thought: I've got it wrong!"' . . . That does happen . . . And I think sometimes speakers just get carried away a bit — they think maths is marvellous and fun, which it is, but you have a lot of other things to do at the same time and you must get it into perspective.

4.3.7 Back in the classroom

Many of the teachers came back to school with ideas which they put into practice in their own classrooms, and they found that they had gained from the course an increased awareness of the mathematical possibilities around them:

> Quite a lot of incidental things . . . things that crop up during the day, like mathematics in the register . . . I think that since the course I've been more aware of things like that, and used them more.

> Just the other week someone brought a sunflower in, and I thought 'Here we go', and so we did lot of estimating, and counting in tens and that sort of thing, and I probably wouldn't have used the full potential of that if

Two-day and three-day courses

> I hadn't been on the course — it might not have occurred to me at the time.
>
> That's the main thing I felt I got from the course — it has made me more aware of using maths in different situations ...
>
> It was therapeutic just to be able to sit back — and the enthusiasm and commitment of the leaders of the groups was infectious ... Now I have widened into more activity ... Also, I was able to look at my classroom as from the outside.

But the pressures of the school environment made it difficult for some of the teachers to keep working on the ideas they gained from the course. And even on a course as generally well regarded as this one, there were some participants who did not hear, or were not able to carry out, the message of the course. Perhaps they were at too early a stage of professional thinking for a course of this type to be effective for them.

> I really think more discussion would have been better, probably a couple more lectures too, and cut out some of the making ... I am just trying to think back to what we did now — we took subtraction at one time I know, and then — I just can't remember what it was we made ... it's so silly.
>
> A lot was said about too much recording going on in infants schools, and you think 'That's right, and I am going to do a lot more incidental and practical maths'. But when you've been back a week or two, you feel the pressure from parents, and from the junior school. I don't know if you found this in other schools?

Reporting back to their colleagues at school, and trying to spread the message of what they had learnt, was a difficult task for the teachers. Even when two teachers from the same school had gone on the course together, the experience of reporting back was a disappointing one:

> We usually report back, and say the things that were going on and the sort of handouts and things like that. I always deviously go back and get a second set of handouts for the school, to put in our resource centre. If you are quick you usually can. People always want to know what you have done — apart from 'what food did you get?' — then they start on 'Who was there, what did you discuss, what did other people discuss that you weren't in on?'
>
> We had a staff meeting very soon afterwards, and we did bring out several of the ideas ...
> Only the main fault is ...
> we didn't ...
> we said about not recording too early, didn't we ...
> really it didn't seem to get over ...
> it doesn't rub off ...
> it doesn't.

But the flavour of the course remained vivid in the minds of many.

Two-day and three-day courses

> I think that kind of course is not to give you any answers, it is just to make you think and to give you ideas. Even if you don't use them immediately, it stimulates your own thinking.

> We are getting away from the old-fashioned idea that maths is sums to a maths that is everything, and finding maths in usual things. Now Polly finds maths in things that you wouldn't think of. We had the topic of food, and she got a tremendous amount of maths from it. She could get maths out of a drawing pin, and I think that is something you need to see. I have more fun out of maths now doing it that way, and seeing the mathematical possibilities in just so many different things, but I do think you need to be made aware of them.

4.3.8 Discussion

This course was concerned with helping teachers to change their styles of teaching, rather than with the communication of content. Its success could be attributed to the determination of the organisers not only to communicate the content of 'maths from everything' but to demonstrate that it could be done — was being done — successfully, by ordinary teachers in ordinary classrooms. This involved detailed long-term planning, and much previous in-service work. It was greatly appreciated by the participants:

> I know Polly did most of the organising for that course, and I think she did exceptionally well. She tried very hard to get a balance, and it wasn't perfect, but compared with other courses that I've been on I think she did extremely well.

> It depends so much on the people who run the courses and the people who ran that course were absolutely marvellous.

Polly's work with the teacher leaders before the course might not have been visible to all the participants, but she had worked to help them to develop to the stage where they were able to sustain their contribution to the course; this was a central feature of the planning of the course. The leaders' confidence was based on secure experience of similar work in their own classrooms, together with detailed planning and preparation over a long period.

Some questions, however, remain after even the best-organised event. First, although the presence on the course of some teachers who already knew that maths was 'marvellous and fun' and who were 'converted to that way of teaching' was a great help in conveying the message of the course to the intended target audience, was this the best use of the precious in-service time of those teachers? Should they rather, perhaps, have gone on a course for the 'converted', which would have inspired them to develop their teaching even further?

Secondly, how far do course organisers prepare their handouts so that they are not only a resource for course members during or after the course but also a summary and a guide for reporting back to

Two-day and three-day courses

colleagues? Should course organisers perhaps encourage the vulture-like tendencies of the teacher who had discovered that you could get a second set of handouts for the school if you were quick? Now that reporting back is becoming a common practice, and a part of school-based INSET, how much training in doing it successfully do teachers need? Perhaps a discussion session towards the end of the course might be used to help teachers to prepare themselves for reporting back.

Finally, in reflecting on this greatly appreciated course, one teacher had a cheering word for all course organisers:

> You go on a course for one thing, and you come back with something else — YOU ALWAYS FIND SOMETHING.

4.4 A residential course for co-ordinators

4.4.1 Structure and monitoring

This course was held at an LEA residential centre, one weekend in spring, and it had an unusual format, runnning from dinner time on Saturday to 4 pm on Monday; this combination of part of a weekend and a school day was much appreciated by the course members. The course leaflet described the course as follows:

> The focus of the course will be on providing assistance to those teachers who have responsibility for mathematics in the 7—11 age range.
>
> Please bring a calculator, a copy of your school's scheme of work in mathematics, copies of textbooks used in your school, and a good supply of coloured pens, rulers, scissors, etc. It would also be helpful to bring examples of children's work, both good and bad.

The researcher attended the course as a participant observer. Because it took place late in the period of the project, the course was not followed up with the participants, but the researcher met one of its members while she was engaged upon another activity, and she took the opportunity of discussing the course with this teacher.

About half the course sessions were led by the LEA Mathematics Adviser, but three teachers from the LEA, a co-ordinator, an advisory teacher and a head teacher, also led sessions, and contributed very considerably to the success of the course. Other local teachers were invited to contribute as group leaders. This represented a change of policy on the part of the Adviser; on previous courses, speakers from outside the LEA had been used to lead course sessions.

4.4.2 Saturday evening

The meeting opened with sherry and dinner; these were followed by the first course session. In this session, the Adviser discussed the

Two-day and three-day courses

writing of a scheme of work; the policy of the LEA was that each school should develop its own scheme of work. A full and helpful handout emphasised that a scheme of work should contain a statement of the school's policy and approach to the teaching of mathematics and its general aims, and should indicate ways of achieving those aims; these should be backed up by detailed descriptions of performance, which teachers could accept as evidence of learning and as 'milestones' along the path of learning and development. These general points were illustrated by extracts from the Cockcroft Report (DES 1982) and *Mathematics 5–11* (DES 1979), and the handout concluded with a checklist of aspects which should be included:

Provision for practical work
Provision for measurement work
Provision for work with shape and space
Graphical work
Logic
Mental mathematics
The use of language in mathematics
Links with other areas (e.g. science)
Provision for high attainers
Provision for low attainers
Ways of fostering good attitudes

The Adviser emphasised the great range of attainment found by the end of the primary years, and pointed out that many schemes did not allow for this:

> ... it is not possible to make global statements about all children ... the aims of the scheme must not be wildly optimistic and unachievable ...

> Many schemes contain glib statements about the four rules of number — they need more expanding ... a good test is how helpful the scheme is to a probationer ...

In discussion, the co-ordinators referred to the need to have precise objectives; some course members seemed to feel insecure about criticism from parents and from colleagues, and a few seemed to wish that objectives were laid down centrally. However, the point was made from the floor that discussion within the school was the best form of in-service, and that discussion improved the teaching in the school, whether or not scheme-writing was actually achieved; doing this work in school made considerable demands on the leadership of the co-ordinator, however. The researcher received the impression that the co-ordinators on the course needed more support before they would be able to settle confidently to the writing of a scheme; perhaps they would have found it helpful to do a small-scale scheme-writing exercise together. However, the handouts and checklist were designed to provide support for the co-ordinator after the course, and to stimulate discussion in staff meetings.

4.4.3 Sunday

The work on Sunday morning opened with a session on investigations led by one of the local teachers; he was a primary teacher and an enthusiast for investigations, who radiated his own enjoyment of this teaching style. He described how he had discovered the investigational approach to mathematics:

> I was one of those who thought that maths is distant, cold, impersonal ... I though it was skills to be memorised ... at secondary school the load gets too much ... it's like if you are learning to play a game and all the time you work on the strokes and you never get to play the game ...
>
> I was lucky to get a second chance to realise I can do it — I can play the game ... not many get the chance ... I was converted ...
>
> I went on a maths diploma course ... the first thing was an investigation ... I panicked — it seemed to be all words — no symbols ... I hadn't realised one could talk about maths ... I saw the answer to my investigation overnight ... realised it was something I had been teaching ... it was a turning point for me ...
>
> Then I heard some lectures ... and realised there was negotiability in maths ... in an investigation we can *create* ...
>
> ... I'm less sure about how to deal with this with other teachers ... when they ask for help about doing topic x tomorrow it's not always productive to suggest some investigations ...

The lecturer distributed two handouts, one on learnable strategies for investigations and the other a list of sources of investigation starting-points. Finally, he set the course members an investigation to carry out themselves, and so the session ended in practical work. All the participants set to work with a will, and immediately became involved in the activity. One more piece of learning was needed, however, for some had not yet learnt to avoid interfering with others' thinking; in the excitement of success, they were apt to tell colleagues what to do.

After coffee, the contributing head teacher talked to the group about leadership and the job of the mathematics co-ordinator. It was, as he saw it, a two-part job; first of all, the co-ordinator needed to prepare policy, and then to help and persuade colleagues to put it into practice:

> ... writing aims does *us* good ... but does it do the children good?

He reminded the group that in the report *Primary Education in England* (DES 1978) HMI had found that 85 per cent of primary schools had a mathematics scheme, but that only 25 per cent of co-ordinators had a strong influence on colleagues. The view in the Services, but not yet that held in education, was that leadership could be developed:

> ... we need to meet both the children's needs and the staff's needs ... you often think you've explained, but they've only heard what they want to support ...

He pointed out that, on different occasions and in different curriculum areas, the co-ordinator had to act in three different roles in relation to other members of staff — as a leader, as a colleague and as a follower. Some colleagues would inevitably be more experienced teachers than the co-ordinator. The speaker's first handout was entitled *Where do we start?* and consisted largely of headings:

Consultation with the Head
Formulate aims and guidelines
Prepare realistic long-term and short-term aims
Build on what you have
Reinforce what is good
Beware the hard sell — prepare your ground
Believe in your product — try to avoid failure
Start with your own class — the best shop window
Keep informed
Communication of ideas
Make sure of your timing
Support the team

Throughout the talk, the importance of building on what was good was stressed, as well as the fact that the co-ordinator's methods needed to be seen to work in his own classroom:

> In a staff meeting ... don't ask for more than people can give ... support the team — people come to you if you are helpful ... delegate — to train for the future ...

After lunch, the Adviser led a practical session on the use of calculators, and then the teachers broke into 'user groups' to discuss the problems of implementing whatever published mathematics scheme was in use in their schools. Among the problems raised in the group that the researcher joined were the organisation of equipment in such a way that it was readily available where it was needed in the school, the value and organisation of group work, ways of avoiding having long queues of children who were in difficulty, and the recording of children's progress. Many of the teachers who spoke were full of enthusiasm for their own organisational methods, but some found it difficult, as a consequence, to listen to the ideas of others.

Two-day and three-day courses

4.4.4 Monday

On Monday morning, the third local teacher gave a talk on remedial work in mathematics. His expertise was in the field of learning disability, and his talk concentrated on methods of screening to find those who needed additional help, and the use of activities that would help children's perception and thinking. He also stressed the need for children to work together in small groups, so that they could talk through their mathematics. His talk made a great impression on the teacher with whom the researcher discussed the course:

> ... that lovely talk ... in fact, I went and thanked him ... he did take it into far greater depth than people normally do ... but he was so calm about it all, and I thought he really understands children and how they are struggling ... he didn't talk about maths at all until nearly the end, and yet it was all so relevant to everything ...

Following this, the Adviser led a session on 'Number 7-11', based on the APU's findings on children's understanding of fractions and decimals, and on the SMP in-service kit's list (SMP undated) of discussion points on fractions. The afternoon was spent in discussing, and putting in order of urgency, the Cockcroft Report's list of the tasks of the mathematics co-ordinator, and the course members dispersed at tea-time.

During Monday, some course members made for the researcher a list of their own long-term and short-term priorities as co-ordinators. Extracts from these indicate which aspects of the course had made the greatest impression on the members.

My immediate objectives:
Put up more maths display work in my room.
Get notes from teachers of equipment they have, and what else they would like.
Educate parents!!!
Long-term:
More co-ordination between the classes.
More practical work.
'Allow' games.
Have central resource area, encourage teachers to plan and take, but return when not in use.

Priorities:
Investigations — set up work in own class — possibly introduce a maths club.
Calculators could be included in this.
Remedial work.
Reorganise resources.
On the 'leadership' issue, try to get more dialogue going among the staff.

Two-day and three-day courses

Short-term:
Make new games to vary my teaching of certain concepts.
Offer to share my home-made resources with my colleague.
Hopefully stimulate colleagues to consider teaching of fractions, time, calculators, etc.
Try different approaches to teaching — might get better results.
Long-term:
Provide work and time for remedial maths, using suitable material.
Provide in-service training, firstly on the equipment already available in school, eventually on other aspects.
Choose a suitable commercially available scheme if colleagues still demand one.
Make parents aware of how we teach maths now.

Short term:
Arrange liaison meetings with secondary schools, discuss schemes, refer to Cockcroft Report.
Introduce calculators as a learning tool.
Arrange to have small groups of 5–7 and 7–9 children to help me understand problems of their teachers, and how to help probationary teachers.
Continue meetings with colleagues to work through Teacher's Manual of scheme, discussing the mathematics, and games and resources needed.
Check on equipment in use, ensure that staff know where it is and how to use it.
Long-term:
Reach better understanding with comprehensive school to ensure greater continuity.
Give support to teachers by teaching alongside — if timetable and Head permit.
Introduce some investigative mathematics.

These lists indicate how different co-ordinators focused on different aspects of their task: on the development of their own personal mathematics teaching and of their skills as co-ordinators, on organisational matters such as equipment and resources, and on school-based INSET, based both on formal staff meetings and on less structured attempts to influence their practice.

4.4.5 A follow-up discussion

Soon after the course ended, the researcher obtained the views of one co-ordinator who had attended it:

> I liked the weekend very much ... it was an opportunity to meet people from other parts of the LEA ... as a co-ordinator, I got a great deal from it ... I've gone through this problem of how you actually approach

Two-day and three-day courses

> people to get work done ... How do you talk to your Head .... how do you set about in-service training within your own staffroom ... it's a very knotty problem ... and I felt that the advice Mr Jones gave us was very much to the point ...
>
> The investigation was interesting, but it wasn't new to me ... I've done the maths diploma ... I thought I would rather do this quietly on my own ... I get confused when there are so many people moving about and talking ... I like talking to people about it, but really I want to go away and chew at it by myself ... The satisfaction of doing the investigation was personal ... and the thing about the course was 'How am I going to pass investigations on to other people?'

This co-ordinator's main preoccupation lay in how best to help anxious and indifferent colleagues:

> I'm still feeling my way ... and it isn't something that happens quickly ... it's obviously going to be a slow process ...
>
> ... I talked to the Head ... and really what pushed her in to it is that we were one of the first schools to do our self-appraisal ... so I used that ... I thought 'Here's a reason' ... I showed her all the leaflets we had on the course and said 'Perhaps you'd like to read them' ... she's not really a maths person — she's a language person ...
>
> Our staff here is in a funny situation ... one is being redeployed ... he's been redeployed once already and he has a chip on his shoulder ... and the other upper junior teacher is only here replacing for maternity leave ... and is desperate for a job ... people here don't go on courses a great deal ... One of the teachers ... her face dropped this morning when she heard you were coming ... she felt very threatened ... she sticks rigidly in her own classroom ... and the work is very formal ... very very structured ... she's very inhibited about maths ... if I go near her classroom she feels threatened ...

On Monday afternoon, the discussion of the list of tasks the co-ordinator should carry out had produced some reactions from course members:

> ... but we were obviously completely exhausted by that time ... it was a very intense weekend ... he asked in what order we would put them ... what is the most important thing ... and I shouldn't have said this — I said 'They are all important' ... I think it is very difficult to answer this ... you have to go through the whole range of talking about personal relationships with members of staff, and with your Head ... and the resentment some people have if they think you're trying to shake them out of their rut ...

4.4.6 Discussion

The magnitude and difficulty of the co-ordinator's task must be quite daunting to some of those who undertake it — but there are those who do not even seek the little support that an overstretched LEA Advisory Service can provide; the co-ordinator who discussed the

Two-day and three-day courses

course had found out that attendance had not reached the expected level:

> ... he said the course was under-subscribed ... if he ran another course ... would he get any different teachers?

This comment raises the issue of how an LEA can provide continuing support for its co-ordinators, and whether those co-ordinators who do not voluntarily undertake INSET should be required to do so. Should the LEA concentrate its resources on those co-ordinators who see course attendance as a way to further their own and their colleagues' professional development? When there is so much need, how can the Adviser best deploy his limited resources?

Was this course effective in helping co-ordinators to develop their role? It certainly presented them with a wide conception of the co-ordinator's role, guidance in carrying out the specific task of building a scheme of work, and good experience of one of the newer styles of working in mathematics — investigations.

However, even the volunteer members of this course were somewhat thrown, at the time, by the size of the task of preparing a good scheme of work, and in fact this was not a high priority for those co-ordinators who made lists of their own short-term and long-term objectives. Dealing with resources, dealing with colleagues, and introducing new teaching ideas were more in the forefront of their minds. These tasks, too, will need continuing discussion and support as the co-ordinator's ideas develop. The course seems to have made an excellent start to an enterprise that will need continuing support for a long period.

A notable feature of this course was the quality of the handouts provided. All the speakers provided full, well-thought-out and attractively reproduced handouts. They must have made excellent reference material for the course members to use later and, as one co-ordinator found, they were very useful in helping the Head to see directions in which the school's mathematics could move.

CHAPTER 5

Case studies — short courses mounted by Teachers' Centres

5.1 Introduction

The group of courses described in this chapter were all mounted under the auspices of Teachers' Centres. There must be very many courses such as these taking place at Teachers' Centres throughout the country; Weindling, Reid and Davis (1983), in their study of Teachers' Centres, found that 90 per cent of teachers in the catchment areas of their sample of twelve Teachers' Centres had used the Centre at least once, and 35 per cent had attended one or more courses at the Centre. The most popular type of course at the Teachers' Centres was primary language and reading, and primary mathematics tied for second place for popularity with primary art and craft. It seems very likely, therefore, that very many primary teachers from time to time take part in events such as those which we now describe.

5.2 A three-session course at a Teachers' Centre

5.2.1 Course organisation

The Warden of the Teachers' Centre had asked Mike Howard, a mathematics co-ordinator in a local primary school, to give a course for junior school teachers on 'place value and operations on numbers'. The course was held on three consecutive Wednesday evenings from 4.30 to 5.45, and it was advertised as follows:

> The course will follow a progression from a consideration of the concepts of place value and number analysis to the principles and sub-skills

Short courses mounted by teachers' centres

involved in the four basic operations upon number. The course will involve practical work that is directly applicable to work in the classroom. As this course involves practical work and apparatus, attendance will be limited to the first 25 applications received.

Mike had taken a leading part in work that led to the compilation of the County's Guidelines, and so he was well known to many local teachers, and widely respected. He had previously given a rather similar course to this one, but it had been held in a school and had extended over eight sessions. He planned his three sessions as follows:

1. Infant work
2. Place value
3. Operations

5.2.2 Monitoring

The researcher interviewed Mike before the course started, and he discussed his plans with her:

> I have done some other in-service, but it was the same course as this one, only this is modified because it has to be condensed into three sessions. I don't know why it had to be a three-week course — perhaps teachers don't want to give too many evenings. I would not mind who comes to the course — I would not restrict it to juniors; infant teachers need to know the progression too.
>
> I shall start with a revision of the order of infant activities; the course is aimed at junior teachers, but they often assume too much. I am concerned with developing the concepts needed to think about operations, separately from the skills. Rote learning is an obstacle to understanding.
>
> Primary work can be thought about in three stages: initially there needs to be an emphasis on concepts and not too much numerical work. Secondly, children realise the quantitative aspect — then they need to group, and they need place value and a number system; finally, when they have seen the need and realised the structure, they can apply it to numbers over one hundred.
>
> I will do fractions if there is time — equivalence but not operations. Operations will only come in incidentally. If fractions are well done there is no trouble with decimals; I always do fractions before decimals.

During this interview, Mike presented a strange mixture of confidence and anxiety. He had a carefully-thought-out view of primary mathematics, and he knew exactly what he wanted to get over, even if it was not exactly what the course description said. The researcher's visit, however, provided him with what seemed to be a welcome opportunity to rehearse his presentation, and to try it out before the actual course.

The researcher attended the three sessions of the course. These started with tea, so that the effective time available for each session was reduced to about an hour, and this contributed to Mike's feeling that the course was too short. In the first session, which was attended by fifteen people, he started by discussing the very early stages of mathematical development. Sorting was examined in great detail, at an infant level. He went on to describe the use of Venn and Carroll diagrams and sorting trees. After about half an hour he began to discuss the use of the number track, function machines, Cuisenaire and Multilink to give concrete experience of number. The use of number lines for addition, difference and subtraction was demonstrated. The work had been well prepared, and Mike had copious notes, which he carefully kept to. Unfortunately, he had planned too long a talk, and there was no time for questions or discussion. The session finished with an apology: 'I'm sorry you've had to just sit — next time you will *do*, and it will keep you busy'.

In the second session, the teachers worked in groups at their tables; they grouped paper-clips into tens in boxes, and then grouped the tens into hundreds; sheets ruled in columns were used to systematise the first and second groupings. Mike described the analysis of numbers:

$$98 = 90 + 8$$
$$= 9(10) + 8(1)$$
$$= 98(1)$$

and he emphasised that this analysis was an essential preparation for operations. Abacuses were used, and bases other than ten recommended, so as to cut down the amount of apparatus needed.

In the third session, Multilink apparatus was used by the teachers for work in bases three and four; the pace was very leisurely and the attendance had diminished to ten. There was some muttering at the back, and one teacher was heard to say 'Get a move on . . .'. Probably Mike was again surprised by the passage of time, because he did not gather together the threads of the course at the end.

After the course, the researcher visisted nine of the teachers who had been on the course in their schools, and discussed their impressions with them.

5.2.3 Looking back on the course

Most of the teachers interviewed started off by telling the researcher what a good course it had been, because the lecturer was a practising teacher:

> It was very good — really practical — he had taught it all in classrooms. It lived up to its title — it was about trying to get across number.

> It was Mike Howard doing the course, so I knew it would be extremely good. He concentrated on the early work rather than the later — and how

important it was for children to understand before they put pencil to paper with sums.

However, there was one dissenting voice, even about his realism as a classroom practitioner:

> He seemed remote from the classroom ... he suggested working in groups, but I couldn't do that — you need to be realistic. He read too much from his notes, and didn't involve others.

Nearly all the teachers who attended were junior teachers (the course had been advertised as being for junior teachers) and perhaps the progression towards operations on numbers was too slow for some of them:

> I was interested in it and I enjoyed it, but I found it very drawn out. I agreed totally with his approach. I think the concept that he was putting over was absolutely vital, but I found that for me it wasn't terribly relevant. I've got fourth years, and I've got the third maths set. I think I probably felt frustrated that he went so slowly — the physical taking of all those pieces and putting them there — I really don't feel that I needed that because I totally agreed with everything.

However, this teacher did feel some conflicts between her ideals and her practice in her own teaching;

> I think his whole idea of place value — everything like that — is absolutely vital, but what do you do when you come to the fourth year? I still bash away at place value as much as possible, but I've got to give them a smattering of something else as well, and I cannot afford to spend as much time as I would like. I think it should have been done before, and when you get to the fourth year, with the third maths set, there are some things that you just jolly well have got to get into their heads, whether they understand it or not. I think at least they should be able to add, and multiply, and share and take ... and have a basic knowledge of vulgar fractions, decimal fractions, and simple weighing and capacity, and simple measuring. If they can do the simplest things accurately ... But when to add and subtract — they find the application very difficult ...

One of the teachers blamed herself because she did not find the course very stimulating:

> At 4.30 or 5.30 you are not at your best really, having given out all day, and unless it is very stimulating it is very easy to drift off. In fact, in the maths course I found that extremely easy, because I found it repetitive.

This triggered off the teacher to talk about some of the problems of in-service courses in general:

> The main complaint which we generally have on courses is that some of them — the majority of them — are a waste of time. I've heard people say — like I've said about this course — 'I know all that'. I've heard people say that they didn't feel the lecturer was enthusiastic enough, or that the course was not practical enough, or that discussions tend to be very

dragged out. .... And if the lecturer is a teacher, maybe they forget they are with adults rather than with children, and they put the material across in the same way as they would with children. However, I wouldn't condemn him in any way, because I thought it was very good and practical — and well prepared.

One young teacher, in his second year of teaching, told the researcher that a course needed to be practical to be of value to him. He had gained from this course, because he had tried out in his classroom some ideas he had learnt from the course, such as breaking a number down into its constituent hundreds, tens and units; however, his class were third-year juniors, and he felt that the course did not go far enough for him.

5.2.4 Discussion

Like many courses at Teachers' Centres, this course attracted a number of teachers who were not regular attenders at mathematics courses. Of the nine teachers interviewed, only one went regularly to mathematics courses, and she was the mathematics co-ordinator in her school. Some others went frequently to courses in other subject areas:

> I go to courses for ideas — ideas that work! I've been to lots of language and swimming courses ... one should go in areas where one is weaker — but I don't. I'm chary of putting children off maths, and I haven't been to maths courses because I'm chary of maths myself.

For such teachers, a course needs to be simple and down-to-earth, and, above all, it needs to give practical classroom ideas. Mike was well placed to do this, but, although most of the teachers respected and valued his practical expertise, the course was not seen as a total success. In some ways, Mike misjudged the pace and level of what was needed, and all his careful preparation went into expressing what he knew to be of fundamental importance in mathematics teaching, rather than into the subject matter that had actually been advertised. There was no mistaking Mike's commitment, sincerity and knowledge, and the hard work he had put into his preparation for the course. But although most of the course members respected his practical experience, they were somewhat disappointed — and this can have unfortunate consequences:

> I felt the course was perhaps longer than it need have been; three weeks is a lot to ask of people. If you go three times and you think this could have happened more quickly, and perhaps have been done in two — or even one — it does rather deter you from going next time.

Everyone who is willing to provide in-service work for teachers has to start somewhere. As with most activities, it is unlikely that a beginner will achieve instant success. Mike had no preparation for the mechanics of lecturing, and little experience of it. He was constantly surprised by the rapidity with which time went for him, because there was so much that he wanted to get across; he did not provide a suitable balance of practical activities to break up the hour; he did not allow time for questions or discussion; such practical activities as were carried out did not seem entirely suitable for the group; his handouts would not have helped a teacher to implement his suggestions in the classroom, or to report back to colleagues at school. Being a good co-ordinator does not necessarily equip a teacher to deal with large groups of demanding and critical colleagues during a lecture — these skills need to be learnt. If he could learn how to work with adults, Mike has much to offer, and there are few enough people who are both very knowledgeable about primary mathematics and have daily classroom experience.

How can people like Mike develop course-giving skills? Perhaps the Warden of the Teachers' Centre might involve them in jointly planning and giving courses with himself, so that the more experienced operator can help the newcomer to focus his knowledge and experience of the classroom and of curriculum planning more closely on teachers' perceived needs. A closer analysis of the course advertisement would probably have helped Mike to plan; the advertisement was the work of the warden, and was very attractive. If the warden, or the mathematics Adviser, had had time to work first with Mike and his colleagues in his own school while Mike led their INSET, his induction into providing INSET might have been more comfortably taken and evaluated, before he had to face a larger audience, whose declining numbers over the period of the course must have been painful to him, as must their unenthusiastic response to the slow pace of the work. Indeed, single sessions, rather than a three-session course, can give a beginner more confidence, as well as more time for evaluation and re-planning before he has to tackle another group of teachers.

5.3 A classroom-based course

5.3.1 Organisation of the course

This course was organised by the primary mathematics planning group at a Teachers' Centre. They wished to have some work on classroom organisation in mathematics, and stated:

> Our planning group thought it would be a good idea to go and see first hand how other teachers planned their rooms, rather than setting up a course on planning.

This visiting could not be done in school time, and so three teachers volunteered to leave their classrooms just as they were at the end of a period of mathematics activity, and to invite other teachers to visit them after school. Visits to infant, lower junior and upper junior classrooms were arranged at weekly intervals in June. After the visits, the planning group decided to hold a follow-up meeting in September. The series of visits was widely advertised, and 60 teachers attended the first (infant) visit. The second and third visits were less well-attended, and the follow-up meeting in September was attended only by a small group.

The format of each session consisted of a brief description of the arrangements in the classroom by the class teacher, followed by an opportunity to walk round and look at everything on display and to ask questions of the teacher. There was then a short final period of general questioning, discussion and summing up at the end of the visit. The whole visit lasted about 1½ hours.

5.3.2 *Monitoring*

The researcher attended the visits, and followed up the course by sending a questionnaire to all the participants and to their head teachers, and by visiting two schools to discuss the course with the participants. The questionnaire to participants asked them why they went on the visits; teachers were also requested to list three things they found interesting during the visits. They were asked whether they would find further similar visits helpful, and how the arrangements could be improved in the future. The discussion session at the end of each visit had not been very successful, and participants were asked to speculate about reasons for this. The questionnaire to head teachers asked about the arrangements for reporting back in school, and whether the visits seemed to have been valuable.

5.3.3 *Reasons for attending the visits*

The original impetus for arranging the visits came from the desire of a small group of teachers to observe the organisation in different schools, because they wanted to improve their approaches to group teaching by using a variety of materials and workcards. However, the visits were widely advertised, and many teachers attended who had not been at the original discussions. Moreover, more than half the teachers who attended went to particular sessions rather than attending all three. The reasons for attending given by those who had been at the previous discussions were in accordance with the original intentions:

> to widen my understanding of other schools' approaches, with a view to modifying our own approach;
>
> to see the organisation of other classrooms;

Short courses mounted by teachers' centres

> to see how other teachers tackle mathematics in the classroom;
>
> I was interested to look inside another classroom and perhaps find a new approach that could be useful in my own situation.

However, the additional teachers who came brought other reasons, which often focused on the detail of content rather than the global organisation of the classroom:

> to see what other apparatus and work schemes were being used by others;
>
> to widen my knowledge of available materials, and to help catch up with new published material;
>
> I wanted to have an opportunity to see what other people did and also the standard of work children should be able to do;
>
> it was an opportunity to glean ideas from other teachers.

5.3.4 Points of interest noted during the visits

The points that interested the teachers fell into two groups: points concerned with organisation and those concerned with content. Naturally, however, teachers who went to observe the organisation of the classroom also noted content, and those who were interested in content also noticed the related methodology.

Comments about classroom organisation included interest in the way the upper junior teacher used her own worksheets as an integral part of the organisation of her mixed ability class. In another classroom a teacher noted:

> The emphasis on class teaching among all the varied materials and children working at their own level.

An infant teacher was particularly interested in seeing how mathematics developed in the lower junior years, and the use of a structured mathematics scheme by juniors. There were several comments on the great variety of apparatus on permanent display in one of the rooms, and teachers were interested in the various methods of assessment and record-keeping.

On the content side, several teachers commented on the great number of mathematical games, including home-made games, on show in one room, the variety of activities and the width covered in practical number work, the ways in which mathematics was incorporated in topic work, and the use of a TV programme follow-up. Informal discussion of content with other teachers was valued:

> It was interesting to talk to other teachers about how they dealt with particular subjects — e.g. tens and units work, number recognition and number formation.

(Infants)

Some more controversial notes were struck:

> I wanted to question the formal multiplication tables on the wall [in the infant classroom].

> The keenness we all exhibited to look at children's work and compare it for standard and presentation with our own children.

5.3.5 The organisation of the visits

The informality of the visits, and the opportunity to see other classrooms and teachers in other schools, were greatly appreciated:

> I found the experience very stimulating. New approaches and ideas are always most refreshing.

> Teaching is a very insular occupation, and now that staffing is more difficult, there are fewer opportunities for staff to visit other schools in school time.

> It helps to overcome the insularity of the classroom, where ideas could become stale.

The least successful part of the visits was the general discussion at the end of each session. The teachers proved not to be very willing to contribute to the discussions; their explanations of this centred on their wish not to criticise colleagues whom they felt had shown considerable courage in opening their classrooms and exposing their methods of teaching mathematics:

> Teachers are naturally cautious about making comments which might be misconstrued, appreciating what it means to a teacher to 'expose' her teaching in this way. Many questions were answered informally, as the teachers walked around.

> One is naturally disinclined to criticise a colleague who has so openly displayed a system of working which is apparently successful.

> The discussion might have continued longer if the class teachers had given a fuller explanation of the work they were doing, but this would be a bit unfair to a teacher who was not a maths specialist.

It is worthy of note that the teacher who wanted to question the formal multiplication tables on the wall of the infant classroom replied in the affirmative to the questionnaire's suggestion that:

> People were afraid that comments might be taken as criticism?

Although we may appreciate the sensitivity and tact of these visiting teachers, we must not forget that the visits were intended to raise questions and to provoke professional discussion of the advantages and disadvantages of different styles of classroom organisation and teaching. Other reasons for lack of contributions in discussion centred on the visiting teacher's unwillingness to expose herself in large group discussion, especially as there were sixty people at the infant visit:

> Most questions were asked privately. Many people find it difficult to talk in a very large group.

> I found it more beneficial to ask the teacher concerned about the apparatus, etc., that I was interested in. This would not necessarily interest the others present.

> It was possible to talk to the teachers during the visit, and this was possibly of greater value — while looking at the work and apparatus.

> Sometimes people need time to digest things ... bringing the final discussion session forward from September would have helped.

An experienced leader might have been able to develop the discussion in such a way that it would have benefited both visitors and visited, sharing the variety of experience rather than appearing to criticise what was shown. But on these visits the discussion was not formally led:

> Group discussion needs a leader who will provoke ideas and questions, e.g. 'What did you think about ...?'

It is important that the programme should operate as advertised. The second visit was advertised as being to a lower junior class, but when the visitors arrived, it eventually dawned on them that the class was of third-year juniors:

> The expectation was that we were to see the work of second-year juniors, but they turned out to be third-years. This was established only by asking the class teacher. I thought she had very bright second-years!!

> I feel the material was too advanced for the age-group advertised in the programme — very disappointed, although the teachers concerned had worked very hard.

> I was *very* disappointed that the visit to Lower Green did not deal with lower juniors, as had been advertised. The teachers concerned had spent a lot of time on preparing for our visit and should be congratulated and thanked.

5.3.6 *Professional development from the course*

Two-thirds of the teachers who returned questionnaires about the course had only been on a single visit, and had chosen only to visit the class representing the age-range they themselves taught. Most of these teachers were infant teachers, but many of them taught in primary (5-11) schools. Some teachers see no need for classroom visiting in their own or other schools, and these probably did not attend the course. Others recognise the need, but find it difficult to achieve in their own schools, with their own familiar colleagues — the barrier of the classroom door is still very strong, and this case study shows how anxious teachers are to avoid appearing to criticise the mathematics teaching of others, even when the wish to avoid criticism stifles discussion. Direct confrontation with the problem might help teachers to face their anxiety. The researcher's conversation with one teacher showed this:

> *What would you find helpful to improve your maths teaching?*
> To visit some other classrooms or schools.
> *Where?*
> I don't know.
> *Would you allow another teacher to visit you?*
> Oh ... well ... I suppose so.

However, it may be that some of the teachers who attended a single course session do in fact visit classrooms containing other age-ranges in their own schools, but they need to visit other schools to see classes in their own age-range, because there are not enough comparable classes in their own schools.

Several teachers made appreciative comments about the stimulus provided by the visit:

> I'm interested in the way other schools organise their activities — it stimulates a fresh outlook.

> To gain from other teachers' experience and ideas. To get stimulus.

> It gave an insight into maths teaching in our area.

Discussions and exchange of views with other teachers were also valued, but people often get out of discussion what they themselves bring to the discussion:

> We are all having the same problems, so I don't feel so bad.

> It is useful to hear how other teachers use schemes or apparatus.

> Their enthusiasm renewed me.

In the first case, little positive may have been achieved — and indeed the teacher's unwillingness to change her teaching may have been reinforced. The third teacher went away with a lighter step and a lighter heart — and probably the teachers whom she visited were equally renewed by her interest.

Some of the stimulus provided by these visits could perhaps have been provided as effectively by similar visits within the teachers' own schools. When the researcher visited the school of a teacher who had seen on the visit 'the value of number games and the great variety of these that there are', she saw a great variety of mathematical games being imaginatively used in the classroom of another teacher in the school. However, in some schools, the visiting of other classrooms is not acceptable, as teachers feel that it will produce negative, nonaffirming criticism, rather than the sharing of insights and the tackling of common problems. Similar uncertainties often lead to lack of communication between schools. The infant teacher who was interested in 'seeing how mathematics developed in the junior school' might have seen similar development by visiting her own junior department. However, visiting other schools de-personalises the experience, and enables a teacher to learn from the strong points of what she has seen without having to come to terms with the weak points, as she might in her own school.

Many of the teachers shared their experiences with their colleagues when they returned to school. This was usually done informally, although a few reported back formally to a staff meeting. The questionnaire asked what points had been picked out for sharing, and the answers revealed that it is easier to share specific ideas rather than the general organisation or any new viewpoints gained. Ideas for number games, workcards and worksheets, and methods of recording were mentioned, but some teachers also commented on organisational points:

> The central store and display of mathematical apparatus.
>
> The organisation of infant maths and the use of home-made equipment.

One head reported that there had been informal discussion on how the published workcard scheme seen on the second visit might complement the school's own scheme.

5.3.7 Finally ...

At the next meeting of the primary mathematics planning group, the members reviewed the course:

> People have asked for the same thing in other subjects ...
> They thought it was good ...
> It was a pity there were no children ...
> It would be difficult to have it during the day ...
> It was disappointing that people did not come to all three ...
> The discussion at end of each session did not get going ...
> People did not want to be critical ...
> The people offering would not mind criticism — it can be mutually stimulating ...

Because of the lack of discussion at the visits, the group decided to hold a forum as a follow-up, to focus on the range of classroom organisation seen on the visits. However, the follow-up could not take place until late September, following the June visits, and very few people came.

In repeating such a course, the providers might try to find ways of conveying to teachers that discussion would be valued and regarded as a healthy way of moving forward, rather than as criticism of those who have generously shared their teaching. Discussion leaders, who would have been able to draw out questions about why particular types of organisation were in use, might have helped teachers to focus on aspects of organisation that they could consider putting into practice in their own classrooms. As we saw in the previous case study, teachers who are going to contribute to INSET need to learn new skills — in this case, the needed skill was that of leading discussion among colleagues.

5.4 A single session at a Teachers' Centre

5.4.1 *The meeting and its monitoring*

The Warden of the Teachers' Centre had invited a contributing author of a published primary mathematics scheme — who was also a practising primary teacher — to speak about the scheme; the meeting took place from 4.30 to 5.30 pm one Wednesday afternoon in the autumn. A representative of the publisher was also present, so that teachers could examine the scheme and take publicity leaflets back to their schools. Twenty-two teachers attended, most of them in groups of two or three from a school. The researcher also attended, and after the meeting sent questionnaires to the teachers; five replies were received.

The speaker faced a difficult task in promoting any detailed discussion on his scheme — some teachers knew it well and were using it in their schools; others were seeing it for the first time. As a teacher, the speaker also seemed ambivalent about the morality of a 'hard sell':

> I'm not here to flog the scheme ... but I'll answer questions ...

and so, rather than entering into detail, he spent much of the time in talking about his philosophy of mathematics teaching:

> ... we should talk more maths with children ... we need to learn from the APU, and from the old Nuffield project ... we teach skills and concepts pretty well, but the children fall down on applying them ... children need to learn how to learn ... and how to tackle problems ... and there are some particular skills — estimation, doubling and halving, inverses — like multiplying by ten and dividing by ten ...
>
> ... when you are appraising a scheme, these are things to look for ... and make sure the accent is on place value in the books you use ...
>
> I wish there was more oral maths — more emphasis on approximate answers ... no matter what textbook is used the teacher is always the most valuable asset — so we are always back to people ...

Short courses mounted by teachers' centres

5.4.2 Replies to the questionnaire

The questionnaire asked the teachers why they had chosen to go on the course. Two responses came from schools that were considering changing their schemes:

> We are considering a new maths scheme for the school. I expected to gain an insight into the type of scheme it was, and whether it would be suitable for us.

> We are looking for alternatives to our present scheme — I wanted to look at it and listen to 'the expert'.

One teacher came from a school that already used the scheme, another came beause the fame of the speaker had preceded him, and a young teacher said that she came to keep in touch with what was available:

> We use the scheme in the school, so I was interested to hear any information that might guide us in its use. I expected discussion on any particular aspect of the books that might be problematical.

> I had heard most complimentary reports of the speaker and his realistic attitude to the teaching of mathematics. I expected little more than to listen to an experienced educationalist on his subject, and gain advice and benefit by listening to him, and learn something from his experience and expertise.

> I attended to keep in touch with current schemes that are available, and to confirm any doubts I had about my own scheme or teaching methods. In less than two years' teaching, this was the first in-service course I have attended. I must stress that this was not through lack of interest but practical problems such as lack of transport and other commitments.

The teachers' responses to the course were predictably varied, depending on their own backgrounds and their motives for attending. Two teachers would have liked more detail about the practical use of the scheme:

> I gained a little more insight into the problems encountered when one tries to produce a comprehensive maths scheme. I was disappointed that the talk was not more specific about certain items.

> I thought it was rather general in its content — I would have preferred one area to be highlighted, and to have seen how the scheme developed throughout the school.

Two teachers responded to the speaker's philosophy and personal qualities:

> I gained what I had expected. I found the lecture refreshing, entertaining and fulfilling. Being newly appointed to a post of responsibility, I found it helped me to gain a greater understanding of the general aims of the subject.

The speaker was interesting and stimulating, and reinforced many of my own views on the teaching of maths, and its difficulties and problems.

The young teacher gained some reassurance from what was said, and would have liked another course of the same type, but with several contributions from different teachers:

> I found that a lot of the points made were things that I was very aware of, and to some extent included in my teaching of mathematics. I wasn't surprised at the suggestions that were made, and indeed felt relieved that I was tackling them in the classroom ... I would like more contributions from other teachers who could explain how they see maths in the school curriculum, the methods of classroom organisation that they use, and the resources that are available in their schools — equipment and schemes of work.

Finally, the teachers were asked to describe any good and bad experiences that had happened to them in mathematics INSET, and any particular in-service work they felt in need of. This elicited one description of a 'bad' course which a teacher had previously attended in another part of the country, and one cry straight from the heart:

> Not a 'good' course; although it was designed for upper primary teachers, it dealt *entirely* with infant number, and did not relate to the particular needs of the heads and staff who attended it, except to a minor degree.

> Any talks and discussions with people who have had prolonged experience of teaching maths and have learnt by experience, and can pass on ideas and tips which have proved useful; constructive help from people who understand the difficulties and frustrations of teaching maths in a practical way in a large class of infants, because however good the apparatus or scheme of work, it falls down if the children do not have sufficient guidance and supervision from the class teacher.

5.4.3 Discussion

Teachers who attend one-off lectures at Teachers' Centres are perhaps more various in their professional development than those who turn up at any other form of INSET. Meetings such as these require no on-going commitment, and so attract some teachers who find it difficult, or are too anxious about mathematics, to attend anything more demanding. These meetings also provide opportunities for new teachers to dip their toes into the pool of INSET. Habitual course-goers are also to be found at these events, but at least they know what to expect, and are perhaps more able to take what they can from the course. We should not underestimate the value of a generalised shot in the arm from an enthusiast — even if it was not quite what was advertised. And, as on so many occasions when teachers describe their INSET experiences, the usefulness of reassurance that the teacher is 'doing some things right' in mathematics came through in the young teacher's report of this meeting.

CHAPTER 6

Case studies — longer courses

6.1 Introduction

This chapter contains case studies of two much more substantial courses than those described in the last two chapters. Both courses took place in higher education institutions, and the course providers were teams of members of staff of these institutions. Colleges are the usual venues for longer courses such as these, which are beyond the resources of an LEA or Teachers' Centre to run for itself. One course consisted of full days at intervals over a two-term period, and the other was a one-term full-time course. Courses of this length can provide much more substantial input, and these two courses were the only courses monitored which overtly attempted to improve the participants' knowledge of mathematics, as well as helping to develop professional skills. One course was intended for co-ordinators or intending co-ordinators and the other was aimed at teachers at an earlier stage of professional development.

6.2 An intermittent course for mathematics co-ordinators

6.2.1 Organisation of the course

This course was intended for mathematics co-ordinators in primary schools in two neighbouring LEAs. It was provided in a higher education institution, which was also involved in the initial training of primary teachers. The course was run at the request of the two LEAs, each of which nominated ten teachers to it and seconded them intermittently at intervals over a period of two terms. The publicity material included the following statements:

> The course aims to equip primary school teachers for the co-ordinator's role in their own schools, and is organised for the X and Y LEAs.
>
> Supply cover is provided when necessary.

The course had been running for two years, and was the successor to a similar course at another college which had closed. The course was made up of a combination of single days and blocks of three or four days, scattered through the autumn and spring terms. The pattern of the blocks and single days had varied each year, depending on the availability of the course tutors, who were two members of staff of the higher education institution. In some previous years, a third tutor had taken part. Members of previous courses had criticised the positioning of a block of time that was too near to Christmas to make release from school easy, but it had not been possible to change this. Each day was divided into three sessions, two in the morning and one in the afternoon; the middle session was left free for the teachers' personal exploration of resources and for use of the library. The other two sessions were sometimes given to the same topic and sometimes to different topics. No specific work outside attendance time (9.30–3.20) was required of course members.

The course content was listed in the pre-course publicity material as covering a wide range of topics, including:

- aims and objectives in mathematics related to LEA guidelines and school schemes of work;
- teaching methods, including problem-solving and discovery methods;
- a detailed examination of some commercial schemes and their back-up resources;
- assessment and evaluation;
- continuity between infant, junior and secondary schools;
- mathematics across the curriculum;
- teaching materials: individual course work related to developing personal mathematical knowledge;
- organising school-based in-service work;
- stages in concept formation;
- the role of the mathematics co-ordinator.

The course tutors had contrasting experience and styles; Mrs Winterbottom had long experience of teacher training, while her much

Longer courses

younger colleague, Simon, was a newcomer with recent and varied school experience; he had taught in primary and secondary schools and had worked with adult students before joining the college. Very occasionally, visiting lecturers contributed to the course, and at the end of the second term each course member gave a short talk on some aspect of the mathematics curriculum, prepared in conjunction with the other course members from his own LEA. The course also made several visits to schools and to Teachers' Centres.

6.2.2 Monitoring

The researcher monitored the course by attending, as a participant observer, a considerable number of the course sessions and visits. She also visited five of the course members in their schools and had discussions with them and, where possible, with their Heads. The course tutors made available the evaluations made by teachers who had attended the previous (very similar) course, and the researcher had formal talks with the two course tutors together, as well as joining in many informal chats during the course.

6.2.3 The course members

The course had been requested by the two sponsoring LEAs, and it was advertised within these Authorities as intended to equip teachers for the mathematics co-ordinator's role in their own schools; although some course members were already co-ordinators, by no means all of them were:

> They make their applications but the applications are vetted by the Adviser so it's definitely part of the LEA's development plan. Not very many of them are already co-ordinators. The majority are people that the LEAs are obviously thinking might apply for co-ordinator's posts. Quite a number of these people have been concerned with working parties, involved in Guidelines.
>
> (Course tutor)

> Mr Martin came in and asked would I like to go on a maths course. He said he thought I was the best person to go from this school. It was the second day of the course before I realised it was for co-ordinators. Most people on the course had a scale post, but I did not even have any ambition for a post. I came back with enthusiasm, and told people about it at a staff meeting; the staff were then aware that they could ask me for help. I don't co-ordinate maths, I just help anyone who comes and asks.
>
> (Teacher)

This teacher's Head had offered her the co-ordinator's post, but she did not have enough confidence to take it, and another teacher was appointed. However, she enjoyed the course and gained from it.

Perhaps not all the course members were willing volunteers. As one member of the course confided to the researcher:

> Another thing is that there are people on this course who are not desperately keen on maths — if you talk to them privately — they have been told that they've got to do maths.
> *And now they've been sent on the course to try and make them a better co-ordinator?*
> Yes, that's right.
>
> Some of the people on the course have said 'Oh, I can't do that, I'm no good at maths.' If they are no good at maths, then what the hell are they doing here?

6.2.4 The course content

The range of background among the members made it very difficult for the tutors to plan a suitable course. They decided to devote much of the ten days of attendance in the autumn term to mathematical content: graphs, sets, number, measurement, fractions, shape, enlargement and similarity, ratio and proportion. There was also some work on investigations and on mathematical games and puzzles. Towards the end of the term, a day was spent on the study of commercially published mathematics schemes, and structural apparatus was considered. On the final day of the first term, three members of the previous year's course talked to the group about the role of the mathematics co-ordinator.

The second term must have flashed by: a day was spent on the development of mathematical topics through the primary years, a day on the extremes of ability, and a third day on LEA Guidelines. Four days were spent in visiting schools and the Teachers' Centres belonging to the two LEAs, and then each participant gave a talk to the group. There was a little further work on investigations, the course was evaluated, and it was time for the members to disperse.

The course content received a mixed reception from the participants. Some of them were enthusiastic, or retained a generally favourable impression:

> I have changed my way of teaching mathematics. I feel more confident and this has affected the children. When I prepare my maths work, I now try to look ahead and ask the question 'Why are we doing this?'

> The course is good. It helps in the job and it's good meeting people with the same experience and the same school problems — it makes me feel they can be solved because others have done so.

> The course was good; we could handle apparatus. I had read about some good ideas in books, but I had never done them. It made one think as the children have to.

> It has made me appreciate in depth and more fully the necessity for practical maths experience before anything is recorded.

Longer courses

Other course members were less certain, and these were probably the teachers who were furthest on in their own professional development. Indeed, of the last four speakers, the second was only in her second year of teaching, but had already been appointed mathematics co-ordinator; the third was the very unconfident teacher quoted above. Some other participants found that the emphasis of the work was not quite what they had expected, nor what they believed they needed.

> I am disappointed in the course — I thought it would have a lot of new ideas, it would be about the role of the maths co-ordinator, about how to help lead the staff, about how to co-ordinate the maths scheme, but it's not. In our school we need school-based in-service and I need to know how to run it — how it can be done so that it helps the staff.

> Things that need to be altered: a bit less of the basic knowledge of mathematics. I certainly feel we could have done with talking about various pieces of equipment; we never got to comparing Osmiroid maths equipment, say, with somebody else's. We talked about the various schemes — well, we did and we didn't, we never really got down to the Scottish, so there's been a lot left out.

> I wanted to talk more about the methods of teaching that can be used for maths. I would have liked to look at schemes in detail — spend a whole day on one, with some good input on what to look for; we need help in looking at a scheme. There was nothing on timetabling within a class — about the integrated day, or maths all at the same time, or teaching styles . . .

> Not enough on the development of a topic. Management in general was missing. The section on the role of the maths co-ordinator was poor. I expected more on how to arrange the school and on approaches; how to help members of staff and how to manage.

The course tutors' planning was based on a pattern which they used for other substantial courses in primary mathematics, such as the Mathematical Association's Diploma for teachers of the 5–13 age-range; in this much longer course, half the time is spent on mathematics and the other half on mathematical education. In the course for co-ordinators, the tutors wished first of all to ensure that the participants had thought about the mathematical content that can be taught in the primary school, and then to use this common experience as a springboard for consideration of various facets of the co-ordinator's work. Group cohesion was also important, so that teachers could discuss their problems without feeling threatened. In discussion, the tutors talked about their ideas.

> They haven't got as long as the Diploma course, so we don't go into anything like the same detail, but we are basically doing much the same sorts of things. It is so short, and they know they are only getting a look at the surface really.

Longer courses

> I suppose the first term we are more or less consolidating a joint position so that we've got common experience on which to build. And then in the second term we are very much more concerned about what they are going to do when they leave, and we have a lot more visits to centres of one kind or another and to resources that are available and visits to schools. We are beginning to think about what's going to happen when they start their co-ordinator's work.
> *Is there anything explicit on their communication with other teachers, or have they got to find their own salvation in handling that?*
> I don't feel competent to help in this way. If we could find a way of looking at this, I would be very happy to incorporate it. There are people in college — in the social sciences — who might be able to help, but I think if they were to be involved they would probably want considerably longer than we can spare them time for. And they wouldn't necessarily relate it to the fears and pressures that people find in maths.

The tutors relied to a considerable extent on the teachers' learning about management and the co-ordinator's role from each other; however, the experience of giving a talk to the group gave course members a useful insight into the problems of making a presentation to other teachers:

> I think one thing that comes over is that the experience of discussing with other teachers is absolutely vital to the course and they spend quite a lot of time talking together.
>
> (Course tutor)

> We had to do a presentation ... I was very nervous ... adult-to-adult isn't easy.
>
> (Teacher)

However, three members of the previous course gave talks on the work of the co-ordinator, and this enabled some of the course members to express their worries as co-ordinators. These worries seemed to centre on relationships with colleagues:

> *Did you have no teacher problems?*
> Yes, there were some.
> *We would have liked to have heard them.*
> I have one teacher who may not adopt Nuffield.

The problem of cramming a quart into a pint pot — or even a litre into a centilitre container — was appreciated by some course members:

> The course is not really long enough; they are trying to do what would need a term or perhaps half a term — you get the taste of something. It's rather like when you sit down at a lecture and the person says to you 'Right, now I am going to do with you in 10 minutes what would take a class an hour to do.' And at the end of it you wish you had not done it in 10 minutes — you would rather have done it in an hour.

Longer courses

6.2.5 The mathematics element

> We have a mixed ability group at this table — some have felt they have been left struggling — there have been many new aspects in mathematics too quickly.

So said one of the teachers at the review of the course at the end of the first term. They were indeed a group of very mixed attainment, ranging from a teacher who had studied A-level mathematics to some who had given up well before O-level. Their comments about the mathematical content indicate the range of their views:

> I have learnt a great deal from the aspects of maths covered and feel it cannot be improved.

> It reinforced my knowledge of maths — I found that one or two of the lectures gave me a different insight into things; it made me understand things that I had thought I understood fairly well. Some co-ordinators do need the maths content, but I know some of the group are disgruntled if they know the maths.

> Some activities were taken to far too high a level to really be of use.

> Some mathematical ideas were dealt with at too high a level, and while I enjoyed listening I don't know how much I have retained and whether I am capable of using it in my class.

It would not seem that there is much content dealing with the understanding of fundamental mathematical ideas which is appropriate for the whole of a group that includes such a diverse range of teachers. The difficulty is very similar to that of coping with the 'seven-year difference' in a primary class. The problem deserves to be discussed with the course members, and ways of coping with it certainly need to be explored, for it is a problem that plagues many in-service courses in primary mathematics.

Some teachers wished immediately to relate the mathematics sessions to what they could do with children the next day. On the first day, the teachers measured each other in a variety of ways and recorded what they had found graphically. The tutor's injunction, 'Notice the value of diverse group activity which contributes to the total', was greeted with: 'But how much of this can one give to fourth-year juniors?'. The questioner's sceptical tone seemed to suggest that he thought the activity was intended to be given to children as it stood: 'Could average fourth-years work out scale?'. As they left at the end of the day, some teachers were heard saying: 'I'm going to do this measuring in school next week'. Although the wish to try out new activities is admirable, the conversation suggests that teachers need to be clear about two things: (i) when they are doing work that is intended to be suitable for children, and (ii) when they are doing work intended to deepen their own knowledge at an adult level. Again, later in the course, participants commented that the number ideas were good — they gave you something to take back to school.

But the fundamental difference remained between the course members who found the mathematics content helpful and those who wanted to concentrate on the work of the co-ordinator. One teacher proposed a solution to this problem, which might be considered:

> A parallel course, so that this course is not concerned with just updating one's own maths. The parallel course would be concerned with just maths. This course is for maths co-ordinators.

On the other hand, would teachers who are seeking promotion come forward for a course whose description might suggest that their own knowledge of mathematics is inadequate for the jobs they are seeking?

6.2.6 Group dynamics and teaching style

One unforeseen phenomenon that the researcher observed was the nature of the interaction during the sessions, both between course members themselves and with tutors. At times, there seemed to be something almost akin to a discipline problem:

> I think a group of teachers together — you saw what we were like — the footballers got together. We are like a very bad class — you are talking to people who know all the tricks, because they have had them in the classroom.

This problem was no doubt at its most difficult when the participants did not see the work as being immediately relevant, or when the presentation was less than gripping. In contrast:

> For example, the day we had the lecture on gifted children and the low attainers everyone was very keen; they listened and they learnt a lot. Now, that was something they had themselves experienced on the job.

Problems of group dynamics did not only manifest themselves during sessions when the course tutors made presentations. On one occasion, when teachers were reporting back after group work, the researcher noted:

> There was a great lack of patience among one group who are not able to listen to the others. This was manifested by them today, and has been true at many levels throughout the course. This does not augur well for co-ordinators; they have to be able to consider others' problems too.

Classes of teachers sometimes seem to revert to behaviour more typical of classes of children; perhaps the bahaviour of some of the group members on this course could be explained by the fact that they were not volunteers. Out of school time, if the offering does not interest a teacher, she can drop out of the course, but when she has been seconded by the LEA this option is scarcely open to her. On this course, different members reacted differently to the problem, but it was not solved:

> So you get the people who will make the funny comments and you get the people who turn round and look at you, who think it's terrible if you even talk.

> Yes, there was nothing destructive in that situation, but some people feel there should be a certain code of practice — a certain way of behaving. In fact, some of the teachers were expecting to behave like children and not like adults who don't interject.

The roles and the responsibilities of all the participants cause problems that trouble tutors as well as students. Teachers are quick to notice the methods and attitudes of the tutors:

> I liked Mrs Winterbottom — tremendous character — really knows her maths. Simon used first-year student techniques. He is a very nice character and I think he has a lot to offer, but I think the topic he was given at the beginning got him off to a wrong start.

> There is much repetition from college, and we are treated like children at times — or perhaps like first-year students.

> I think we should have been given something explicit to do during the free sessions. The library was not enough: there is a need for the lecturers to be directive.

> The outside lecturer who came in — low attainers — he was marvellous. He's attached to the psychology department at the college. That lecture definitely raised everyone's spirits, and they came out quite enthused. Some of his handouts were very good — it's not something one expects; well, I expect it — if I go on a course I expect to come away with something.

The course members had an obvious affection and respect for Mrs Winterbottom, but they were still able to disconcert her when some of them found one of her more imaginative teaching situations comic:

> And the people on my table laughed and thought it was very funny, but lots of people on other tables didn't think it was funny at all. She took it very well.

In truth, as one of the teachers said, 'Adult-to-adult isn't easy'. Simon was a comparative newcomer to in-service education, and was still developing his style:

> It was new to me — I came into the college to work on these courses, and I felt very nervous ... but it appears to have gone reasonably well. I learnt a tremendous amount last year, and this year I did it rather differently. I know I don't find it easy to lecture in a traditional way; I don't work like that as a person, and I think working with a group this size I found it much more difficult. So I tended to split them up into

Longer courses

groups and give them questions to discuss and then get back together and feed back.

It may be that this particular teaching style gives teachers the impression that the work is not always thoroughly well-prepared:

> A lecturer ought — like a good teacher — to have everything really planned, and I think some of the lectures were not planned or might have been planned but the content was poor and it led to boredom.
> (Teacher)

Rudduck (1981) has commented on the value that course members attach to handouts, and this is a point for consideration by those whose teaching style involves drawing out points from the course members themselves. Lack of handouts may be equated with lack of preparation; a summary handout, either prepared in advance or distributed at the next meeting, may help to give course members the impression that tutors have something of their own to offer and that they do not merely drop in to see what course members can tell them. The difficulty of note-taking weighed on some teachers' minds:

> Students are different — they do it all the time — but we haven't taken notes for years.

> It's very hard to look and watch and take it in and take adequate notes at the same time.

The last comment was made about the mathematics sessions, but it is even more difficult to record the salient points of a discussion session for future pondering. And if the teacher has not retained a clear structure from a mathematical education session, reporting back to colleagues in school must be extremely difficult, and they may gain a somewhat garbled impression of the course.

6.2.7 *Course structure*

Course members greatly preferred the three- or four-day blocks of time in college to the single days; blocks of time gave opportunities for concentration without the teachers' worrying about what they had to do in school the next day. However, their absence did cause some problems back at school.

> The groups of days were disruptive to the schools. (LEA Adviser)

> I thought it was tremendous — you go for a week; you get into the routine of travelling there, which is a trauma in itself sometimes. So you are geared up — next week I am off to college and I know what time to leave. So you are all geared up to go there and then you start getting into the library — using it. I've used the library quite a lot now, but I didn't to begin with. It's only when you are there for a couple of days that you

Longer courses

> begin to go and sit in the library and read — and I think we suddenly became students again.
>
> (Course member)

This opportunity to become students again was certainly the intention of the course tutors. Members of previous courses had asked for free time during the course days, so that they would have time to read and work during the course sessions; consequently, the course tutors were very anxious not to direct what the teachers did in the free session in the second half of each morning.

> The session in the middle we have left free. This is deliberate policy because we felt that each person coming is going to have very different views on what he is going to get out of the course and will want to look at the resources in the college in different ways. We've got to give time for that, and on the whole they are quite pleased about it. But interestingly enough the teachers have been saying why don't you direct us more about what to do during that time.
>
> (Course tutor)

Perhaps the fact that the teachers were not required to do specific pieces of work other than prepare their talks for the end of the course was responsible for the repeated requests of members of this course not to have a 'free' session during the day.

> Much more should have been demanded of us. The gap of time at the end of the morning should have been more constructively planned and more expected of us. The talks at the end forced us to draw our ideas together and so they were good.

> We have wasted two hours every day. I appreciate that the staff could not be available, but we could have been left something positive to do.

Judging by these teachers' comments, perhaps they would have felt more satisfied if they had been expected to produce assignments contributing to a Certificate of Further Professional Study or other qualification which would have been marketable when they applied for co-ordinators' posts. The fact that the teachers' applications were vetted by their LEAs did not automatically ensure that the participants were committed to mathematics and would contribute to its development in the LEA on their return.

Another (again perhaps avoidable) cause of criticism by course members was the placing of the mathematics sessions before the work which more specifically related to the role of the mathematics co-ordinator. Certainly, as we have seen earlier, the tutors had considered this point, and had made a reasoned decision to structure the course in the way they did. However, the course members did not altogether grasp the reasons. As one of them told the researcher during the second term:

> Improving the way we teach maths I think is very good, but too much

time was spent on that aspect. More should have been spent on comparing attainment check-lists: I think we should have visited maths centres sooner in the course rather than now. It's almost like a poor primary syllabus — you do the basics first and then we will go on to the nice things.

Indeed, probably a reason for many complaints was lack of communication. The course certainly had been carefully thought out in advance, and course members had received a programme for all the sessions, but they claimed that they did not know *why* they were doing the things they were doing.

It would have helped to have had some structure to apply to schemes. The major factor is time; we don't get as much time as we would like, so structure is needed. Also we don't really understand the reasons for what we are doing on the course; we would like to have some reasons, e.g. your objectives.

6.2.8 Discussion

When participants are asked to evaluate a course, it is inevitable that grumbles and suggestions for improvement come to the fore. Probably the basic problem of this course was that it would be impossible both to update the teacher's own mathematical knowledge and to train a co-ordinator in a course of this type and of this length. The course providers had to make difficult choices between skimming the surface of many topics and covering a smaller number in depth. They also had to resolve the appropriate balance between discussing the mathematical content of the primary curriculum, study of the teaching and learning of mathematics, and management skills.

There was a considerable mismatch between the pre-course publicity, with its suggestion that the course would deal with a broad range of topics of interest to mathematics co-ordinators, and the actual emphasis, throughout the first term, on helping teachers to develop their own mathematics, and to strengthen their own mathematics teaching. Some topics which were advertised, and in which a number of course members were interested, such as the role of the co-ordinator, how to organise school-based INSET, and the examination of schemes and apparatus, received comparatively scant treatment and were positioned late in the course; by this time, some disillusionment had set in, so these topics were perhaps not appreciated as much as they might have been.

However, for some participants, it was exactly the right course. As one member wrote in her course evaluation:

Generally — most informative
— re-enthused.
Can I do it again next year, please?

Longer courses

Teachers who took this view were clear about their need for the mathematical content; they were at an early stage of professional thinking, when they were still concerned with establishing their own personal teaching styles.

For other teachers, who were further on in their professional development, a course with a greater emphasis on the range of mathematics teaching throughout the primary school, liaison, co-ordination, and the basis for choice of content and resources, would have suited their needs better. The course these teachers needed was probably not available within their area, but it is not always the case that the wrong course is the next best thing to the right course, especially if there are other teachers for whom it is the right course and who would benefit much more from it. This type of problem will continue to occur until INSET is available in greater quantity and variety, and until opportunities for such training are expected to occur more frequently during a teacher's career.

Perhaps the advisers from the LEAs concerned were not totally clear about the type of course they were asking for and the range of teachers who might best profit from it. Certainly, some of the teachers who were encouraged by their Advisers to attend found that it was not quite what they had expected. The result of the method of selection, too, seems to have been the gathering together of a group of teachers at such diverse stages and with such different needs that satisfying them all presented an almost insoluble problem to the course tutors.

The tutors, as well, were limited in what they could confidently offer; their position within a college meant that they had less experience of how teachers could develop in order to fill the role of mathematics co-ordinator than they had of other aspects of primary mathematics. At that time, the appointment of a mathematics co-ordinator was an innovative idea in primary schools, and there was little expertise anywhere that could have helped newly appointed co-ordinators towards the effective management of curricular change within their schools. It is always the case that when educational innovation occurs, courses are needed to help teachers implement it, but these are the courses where it is most difficult to satisfy all the participants. However, the course tutors had been requested by the LEAs to run a course 'to equip primary school teachers for the co-ordinator's role in their own schools', and they did their best to provide activities they regarded as suitable. Indeed, the activities were recognised by about half the participants as appropriate to their needs; they might have been regarded as suitable by the remainder if they had not been concentrated so early in the course.

This case-study makes clear a few of the problems caused by the lack of overall planning needed to support the professional development of teachers. The same course cannot be expected to satisfy

Longer courses

teachers at widely different stages of professional development. Moreover, little guidance is available that would help teachers to appreciate what forms of INSET and types of course would be best suited to their own needs at any particular time.

6.3 A one-term full-time course

6.3.1 *Course organisation*

This intensive one-term course took place in an institution of higher education, which includes among its other activities the initial training of primary teachers. The course enjoys considerable support from the LEA in which it is situated; it has been offered for several years and the course tutors have developed a detailed outline of its aims and procedures. The aims are stated as follows:

> The general aim is to improve teachers' understanding of mathematics and their ability to teach it; more specifically:
>
> - to explore mathematical concepts and their development, in particular the concepts of number, shape and measurement;
> - to provide an understanding of more recent topics such as probability, co-ordinates, and computing;
> - to suggest methods, activities, apparatus and resource books suitable for various ages and topics and to discuss their use;
> - to discuss various forms of class organisation for mathematical activity;
> - to explore ways in which mathematics arises in in other activities, e.g. in project work and environmental work;
> - to give overall coherence to the mathematics taught in primary schools.

The course consists largely of combined lectures and practical sessions, with weekly visits to a range of selected schools. There are also a few talks from outside speakers. The course is very intensive and the lectures/practical sessions alone exceed 150 hours. The timetable gives regularity, but is flexible enough to permit of change when this is desirable. The teaching team contains five tutors; a room which is well-equipped for practical work is allocated to the course for the whole term. The syllabus is set down in terms of mathematical content; it has varied little from year to year and contains substantial sections on numbers, measurement, shape and graphical representation. The work also includes introductions to computing, matrices, algebraic ideas and simple linear programming. There is no formal

99

Longer courses

assessment, but course members are required to undertake a number of activities, such as:
- producing results, conclusions and comments on practical work;
- presenting project work, both verbally to the group and in writing;
- preparing teaching aids and materials;
- mounting an exhibition of their work;
- constructing an appropriate mathematics syllabus for their own school.

On the last day of the course, the teachers' Heads and colleagues are invited to see the exhibition of their work; during the term following the course, the teachers are invited back for a half-day session, to share the experiences they had on returning to school and to show some of the work they have done with children.

6.3.2 Monitoring

This course was monitored in considerable detail. The researcher attended the course as a participant observer for at least half a day each week, and took part in most types of course sessions. On the first day of the course, the participants were asked to write in a few lines why they had come on the course, and what they hoped to gain from it; on the last day, they completed detailed evaluation forms of a type that had been used for each course, and also held discussions with the course tutor, Janet. Course members decided they did not wish the researcher to be present for this session, although afterwards they said it would have been all right. The researcher also held detailed discussions with Janet and with the other tutors, both before the course and during it. After the end of the course, the researcher visited thirteen of the nineteen teachers in their schools, and discussed the course with them and their Heads; she also attended the follow-up session in the college.

6.3.3 When the course began

The course members gathered at the beginning of term with a varied set of hopes and expectations. Some were looking for a greater understanding of the development of mathematics in the primary years, or wished to consolidate their own knowledge:

> I would like to know why we teach what we do in infant school and to see the topics developed at junior level. I also find maths fascinating, and I expect to be re-stimulated.

> The reason I came was generally to give me more insight into maths throughout the entire primary school.
>
> I came on this course to improve my understanding of the mathematical needs of children. I am going to take a closer look at what I am doing, why I am doing it, and in what order it should be presented.
>
> To consolidate my ideas and obtain a realistic and practical progression for the child. I am educating myself in maths, which I have missed out on in the past.

Several had trained for a different age-range from that which they were now teaching:

> I was infant/junior trained but we did mainly infants. I have now been teaching in a junior school for five years. Therefore I feel I need more junior experience in maths.
>
> I came because of not being primary trained; I felt I was too formal in my teaching of maths and hoped the course would broaden my outlook on it.

Some felt that the course would improve their promotion prospects, and others that a break from the daily round of teaching would give them a fresh stimulus:

> Although a lot of discussion of maths has gone on in school, I felt that my knowledge was very limited and so the course would help me. Also I was stale and needed relief from class and school.
>
> I hope to feel re-stimulated by a break from children for the first time for several years.
>
> I was applying for deputy headships and being unsuccessful at interview, so I thought a course would help. I chose a maths course because I am interested in maths and wanted help with more modern topics, e.g. probability, co-ordinates and environmental maths.

The course tutors also described their own hopes for the course. This was the fifth time they had run it, and they had evaluated each course; consequently, their aims related closely to the aspirations that the teachers had expressed. They would concentrate on developing the teachers' attitudes, in order to give them confidence and enthusiasm. They would also work for an understanding of progression and integration in mathematics, and towards better classroom communication with children:

> I hope that by the end of my part of the course, which is on number, they will be feeling more confident about the material they teach, and about ways of teaching it. They should have more understanding of number, and this understanding will raise their expectation of understanding for children. Some of them will have developed an overview of number ideas, and so will feel more able to initiate, assess, and develop individual approaches in their classroom and possibly throughout the school.

Longer courses

> They will have used, and been asked to reflect on, a variety of materials and apparatus — and they should be better able to assess, choose and use them effectively. And they should have a greater range of methods to choose from.
>
> They will have experienced a new or renewed enthusiasm for number — the pleasure and delight found in number patterns, the challenge of number games. They will want to share this feeling with children when they return to school.
>
> I'm sure that by the end of the course they will have thought more deeply about maths — in my case through shape and measure — and about learning and teaching it. Their teaching should be affected by this, because they will feel more confident about what they teach, and they will have a better grasp of their material.
>
> I hope the teachers will learn to read maths books as resources for learning, and be able to take responsibility for their own decisions rather than referring to external authorities.

6.3.4 During the course

The researcher attended many course sessions, and described the way in which they were conducted.

> The teachers worked in groups around four large tables, and at first they tended to stay with the same people. Gradually the groups became more flexible, but some continuity remained. I joined a different group each time and undertook the same activities as the teachers. I only provided input when the activity required us to share ideas. I did not take notes or use a tape-recorder during the course sessions, and so I had to rely on my memory for impressions.
>
> The working pattern was usually the same, although different tutors varied the balance. The tutor led an activity, teaching both through direct input and through guided discussion during and after the activities. Many of the activities required teachers to present a display; these provided a basis for questioning and for clarifying the mathematical ideas.

As the course progressed, three attitudes developed: the teachers lost their fears, and became much more open in sharing with one another; the tutors came to an understanding of each teacher's individual needs, so that personal development could be achieved through individual challenges; as the teachers and tutors shared experiences, there was a blurring of the formal tutor-student relationship. The tutors worked hard to create this atmosphere, and the teachers noticed it and commented favourably on it in the follow-up interviews.

The researcher continues her account of the sessions:

> Any course can produce tense situations, because course members are often insecure and worried about possible criticism — and this course was

no exception. For example, one or two people might dominate a discussion; there might be aggressive questioning about an idea, or disagreement about ways to present a display. There was the continual challenge of new ideas. When individuals did not understand something, they lost confidence in themselves; there was sometimes impatience with a tutor when the presentation of a lecture was too fast. There was increasing tiredness as the term progressed; pressure mounted as the time of the exhibition drew near.

Many of these situations were defused naturally, but sometimes the tutors deliberately tried to release the tension through open discussion; for example, the presentation of work was discussed, and when the pressure mounted, plans were modified because the timetable was flexible. The tutors were also very much aware of individuals who needed support because of pressures from the course or other sources.

6.3.5 At the end of the course

At the last session, the course members completed evaluation forms and held a discussion with the course tutor, at which the researcher was not present. The evaluation forms, however, gave a good indication of their views.

The mathematical sessions on number, shape, measure and applications met with almost universal approval, but there was a need for more time for the consideration of apparatus and books. Some 'educational' sessions in previous courses had been criticised, and in this course much of the content had been incorporated into general follow-up discussions; these were found helpful:

> ... glad to have the theory and to think about ideas; it helped to bring the course together.

A feature of the course was the 'link' lectures, relating mathematics to other subjects; these were viewed with some scepticism, because some of the links seemed rather contrived. However, the environmental day received general acclaim. Most course members had appreciated the visits to schools and to teachers, but visiting lecturers varied in effectiveness:

> Some lecturers are better than others at building rapport with the group in a short time.

Some topics requested by previous groups did not easily find an appropriate level. The teachers were divided on the subject of computing (it was then 1980, and few computers had reached primary schools); it was said to be both 'too superficial' and 'not relevant to us'. Testing, assessment and record-keeping had also been requested by previous years, but received a mixed reception.

Useful and enjoyable aspects of the course included discussion and the exchange of ideas, being with adults, working practically, and doing things that children do. Course members said that they had gained individually in:

> enjoyment and confidence;

> opening new dimensions and re-evaluating my classroom approach;

> it was encouraging when I found I was doing some things right;

> I have thought about how to help children understand — by going slower, by rethinking my questioning, by practical work;

> a deeper understanding of how children learn and of mathematical progression, also how to use and assess books and apparatus;

> I used to avoid maths!

Two issues were raised about the balance of the course. First, should it be split into infant and junior groups? The teachers thought there was too much emphasis on one age-group — the junior teachers thought there was too much on infants and vice versa — but many participants valued seeing something of both age-groups. Secondly, there was a strong plea for:

> Time — more study time — but what can go? School visits? Or do we need another term? There is always so much to do.

Finally, there was the exhibition. This event was intended to create a link between course and school, and so to facilitate the teachers' return to school. In addition, it gave an opportunity for the teachers to review the activities of the term. The exhibition included ideas that the teachers had developed in practical sessions as well as their individual projects.

Special invitations had been sent to the Head and staff at each teacher's school, and other teachers could come too. The room was manned by the teachers in turn all afternoon and evening, so that they could explain things to the visitors and welcome their own Head and colleagues. This should have been a happy occasion, but sometimes there seemed to be a certain apprehensiveness as course members showed their own colleagues around. The significance of this became more clear when the researcher visited the teachers' schools later. Some teachers, while they went around the exhibition, were able to talk to their colleagues in a relaxed atmosphere in which there was an opportunity to share enthusiasm and to discuss the implications of ideas. Others gave only apologetic explanations of the work, and had an apparent desire not to appear too knowledgeable and enthusiastic. Some course members had no visitors; some of these knew why their colleagues had not come and understood, but others were hurt by this apparent lack of interest, and it created some apprehension about returning to school.

6.3.6 Re-entry

Re-entry into school was difficult for many of the teachers — both they and their schools had changed while they had been away, and there was a major readjustment to be made. In June, when they had been back in school for just over half a term, they spent half a day in college and talked about their experiences.

> It was a nightmare going back — I wished I'd never been on the course. I felt in such a muddle at first, with too many ideas. I reorganised the class into four groups, and then I had to move to two, to prevent chaos — I haven't found out how to organise it yet. Perhaps it would have been easier to carry on formally, but I wanted to explore; it will be useful for next year.

> My class had been working on homes, so I carried on from there, and we did some work on roads — in the end we did an assembly on it. It's been good to have a term to try things out, but we seem to do *all* maths — but some writing comes out of it as well. I discuss maths each week with the staff; the first week we talked about the exhibition, and everybody wants to try out ideas and show what happened in their class.

> I decided not to make any major changes — my class was in chaos. But I thought I would try a few ideas. We skimmed through lots of ideas, and I found this helpful towards next year.

> I found the first few weeks hard — I couldn't settle. I have the reception class and some children had been new at Christmas, and more came new at Easter. I'm full of everything — enthusiastic about maths — but the others aren't; could anything be done to help the transition back to school? I need a longer holiday to plan what to do, but I'm looking on this term as experimental.

> We've been working on multiplication, and I've spent a lot of time sorting out their understanding. My ideas don't fit with Fletcher any longer, so I'm extracting bits rather than using it page by page.

> My organisation has been the same, but now I'm clearer about how to extend the work — how far it should go. Work has been more fun for me and so the children have enjoyed it too.

> I have allowed them to *talk* more about maths — we have discussed. I have sent them out to do things and *listened* to them. I've gained confidence by trying things out. I let them do a topic on *anything* they liked in maths — they chose their own. But they chose too widely; I should have restricted their scope. I shall try again, but next time I shall have the resources and references ready.

6.3.7 Follow-up in school

The researcher was able to visit most of the teachers in their schools during the next autumn term, to see how they were faring two terms after the end of the course.

Longer courses

In some schools, the attitudes of the Head and the other staff were positive and supportive to the newly returned teacher. Some of the things the teachers had gained most from the course were things that they could do in their own classrooms:

> If a child has a problem I now feel I KNOW how to help. I am thinking differently, and I have a broader outlook on maths. I am using more apparatus, and the children don't feel they are bad if they use it — I don't think any longer that it is only for younger or poorer children. I realise how much children need to talk about maths, just as I did on the course, and so I encourage children to come and talk about their discoveries.
>
> (Teacher)

When the atmosphere of the school made sharing easy, and when the teacher was tactful and informal, it was comparatively easy for her to re-establish herself and make a wider contribution to the school.

> I've had opportunities to share things with my year group, and in general discussion — 'Come on, Susan, what do you think?' — I'm a pretty outgoing verbal person and I can't keep quiet, but it's a staff with a lot of communication, anyway.
>
> (Teacher)

> The staff here discuss a lot, so it was easy to contribute — in fact, I often shared with a friend even during the course and she tried things out then. We have an atmosphere of sharing, and so many of the things I was talking about were tried in all the classrooms.
>
> (Teacher)

> I was worried that the other staff might react against the course, so I started by going into classes and working with the children — in our school the co-ordinators can do that — and it doesn't matter if it fails now when I'm trying things with children. People began to ask for help: they wanted alternative ideas for things that didn't seem to work.
>
> (Teacher)

> Martin's approach is not threatening to other teachers — 'Here is a piece of apparatus I have learnt about; I find it's good . . . I never thought of this myself before.'
>
> (Head)

> Susan's informal contribution in the staffroom is very good and not dogmatic. She has been sensible about making the changes she needed to make, and has gone about it slowly.
>
> (Head)

> One-term secondment is very valuable, as it gives an opportunity to distance oneself and to meet others; however, there can be problems for the school — the supply teacher was poor — and then there is the re-entry problem, and the school needs to offer support. But the long-term effect for the teacher is very great; it provides an opportunity to grow and to develop one's outlook on teaching.
>
> (Head)

In other schools, the teachers who had gone on the course were not able to make much contribution outside their own classrooms, perhaps because they themselves found communication difficult, or because there was little communication in the school as a whole.

> We spend a week on each operation and I feel that they do understand better, now I know it is better to teach operations rather than algorithms. Communication is difficult in the school because of the different buildings — I have not yet given my report on the course.
> (Teacher)

> After the course she did much more maths in her own classroom, but has not yet got a real feel for the integration of the subject. I thought the staff might gain from her experience, and so I have made various arrangements for her to work with different groups of children, trying to get ideas carried back into the classroom, but she seems to have a communication problem.
> (Head)

> James says that he has learnt a lot, but I don't see much change in his classroom. There are still compartments in his mind between subjects, and between areas in maths.
> (Head)

In some schools the teachers' new knowledge was challenged by the need to communicate it to their colleagues, but they were able to rise to the challenge with the support of their Heads and colleagues.

> In the summer it was easy to get straight back — I didn't feel secure in the course environment. When I got back into school, people were prepared to listen, but I didn't yet feel confident enough to talk to others. I kept a very low key at first; I reorganised the maths apparatus because I felt this was needed, I tried to introduce some new cards and extend the system — and I really got on top of understanding the apparatus, and then I felt more confident myself. So then last week I led a staff meeting on Cuisenaire. I chose Cuisenaire because the apparatus is in the school and not being used. We talked about how we could use it in the first year to give a basis for number understanding — for older children it could be used for decimals, measuring and for remedial work.
> (Teacher)

> I am trying to think hard about how to communicate the information to teachers and I am realising that I do not have to tell them what to do, but that it is better to actually do it with the teachers, or do it with the children. I have realised that it is important for the teachers to have activity too.
> (Teacher)

One teacher had a post of responsibility for music before she went on the course, but was expected to take over mathematics when she returned. She found herself having to operate a commercial scheme, which had been chosen and purchased by the school while she was away. This situation was too much for her fragile confidence, and she found it very difficult to cope with.

Longer courses

> In telling the others what I had learnt I had to be careful as they were touchy, so I tried to relate it all to myself, 'I had gone too far, too fast before — I had been too narrow'. I was asking questions about maths before the course, but I used to ask other teachers, thinking they were all doing it right except me. I don't know about advising someone if they need help — I would put the emphasis on what could go on in the classroom and leave them to decide. There are still some things I do not understand and would not want to try.
>
> (Teacher)

> The new scheme was adopted while I was away; it is intended to form the central core. They just got the books, looked at them and bought them for the school. I don't like it all, but obviously nothing is ideal.
>
> (Teacher)

> I am now beginning to ask her things, and she has started to give her views more freely and has relaxed from her fear and anxiety about criticism, but she is still too insecure to go into other people's classrooms.
>
> (Head)

6.3.8 The second teacher from a school

The course had been running for five years, and it served a fairly confined geographical area. Inevitably, some course members came from schools where a colleague had attended the course in a previous year — indeed, in some cases the reason that a teacher came was because a colleague had been on the course earlier:

> One member of staff from my school has already attended the course in the past, and I was impressed by the work and the ideas she brought back.
>
> (Teacher)

These teachers faced different challenges on their return from those met by teachers who were the first participants from their schools. Their schools might already be working on the ideas brought back to school by the previous course member; in such a case, the second teacher would come back into a situation where ideas from the course had already taken root, and so re-entry would be comparatively easy. Alternatively, a teacher might face the embarrassment of returning to a school situation where a colleague's attendance at the course had produced little change in the teaching, either in the individual's own classroom or in the school as a whole; the returning teacher would see this only too clearly.

> I had realised before I came on the course that a lot of what I would learn would be here in the school already. But when I got back to school people wanted to know about it; I felt like a newcomer because I had a different class, but the staff knew what to expect because one of them had been on the course before.
>
> (Teacher)

The main purpose of the course is for the individual and this will help the school, and bring a broadened experience to the school. But its main aim is not for the school — it is not a set rule in this school for the staff to have to communicate about a course.

(Head)

Most of the classroom equipment was probably bought by Barbara, who went on this course about five years ago, but it has not been generally used. Marilyn showed her work-cards at the staff meeting; she felt she was not inhibited, because it was five years since Barbara had been on the course and so some revision was valid.

(Head)

We have had experience of someone going on a one-term language course, and the integration back into school is always difficult; however good the course is, the actual *application* back into the classroom is difficult, and there are always a thousand and one other things to do.

(Head)

Another teacher was saved being placed in the situation of being the second teacher from her school on the course; she was redeployed from her infant school, where the Head had attended the course earlier, to its linked junior school. Thus, she was able to use her new knowledge fruitfully to bring the work in the junior school more closely into line with that which was developing in the infants.

6.3.9 *Discussion*

This course had developed to its present form over a considerable period of time, through close contact with participating teachers and through regular evaluation, so that it was very much in accord with the stated needs of the teachers. It fulfilled most of the expectations that the teachers had when they applied to come on it, and it was always very positively evaluated. The tutors had a close working relationship with the course members, who appreciated their active involvement in the course and their enthusiasm for primary mathematics.

The aims of the course were concerned with 'improving teachers' understanding of mathematics and their ability to teach it' within their own classrooms; these aims were faithfully and successfully implemented. The course did not aim to prepare teachers for leadership in mathematics in their schools, although the tutors did not rule out the possibility that *some* participants might reach this position.

However, when the teachers finished the course and returned to school, almost all their schools had different expectations of them. At the very least, returning course-goers expected to share what they have learnt, and when they have been away for a whole term, they will necessarily have learnt a great deal. It is probably easier to share the results of a single-session course on one topic than it is to share the attitude changes and the new ways of thinking that this course provided for many of the teachers. Many of them needed a period of review and reassessment on their return to school, and only gradually could they begin to work out their new ideas in the classroom, and then to share what they had learnt. Indeed, a returning teacher is probably well advised to adapt her teaching gradually, rather than to try to change everything at once, with the inevitable failures and frustrations. This carries the danger, however, that a change not implemented at once may never happen. Perhaps practical and realistic planning for the changes that course members intend to introduce might form part of a course such as this.

Many of the schools expected much more of the teachers than a mere development of their own teaching, and a sharing of knowledge and experience. The expectation of the school was that, on returning, the teacher would take up a leadership role in mathematics. While some of the teachers rose to this challenge, and were able to assume leadership, it had not been the intention of the course tutors, in planning the course and in recruiting course members, to prepare teachers for leadership in the mathematics curriculum.

In an ideal world, teachers at an early stage of their professional development would go on courses that aimed to develop their classroom teaching, and then, some time later, on courses which prepared them for leadership. Because there is not yet enough variety of in-service provision, courses are seen as being 'all things to all men', and it is not surprising if the match is not very close between the aims of the course, the needs of the participant and the expectations of the school. However, much of the success of this course came from its precise focus, and it would not be a useful step forward to try to devise courses that try to be 'all things to all men'.

Going on a course can be a traumatic experience for teachers. They arrive on the course both with high expectations and with some apprehension. Will they understand the mathematics? Will they be able to do the work? Will the course fulfil their expectations? Will their fellow participants and the course tutors be supportive? Not surprisingly, there were some tensions during this course, not least at the end, when the participants, most of whom had come through triumphantly, exhibited their work to their Heads and colleagues — or at least to those Heads and colleagues who made the effort to come to the exhibition. As one teacher said, 'Could anything be done to help the transition back into school?'

Re-entry is a problem that needs consideration in connection with all full-time courses in primary mathematics. Could the college-school links be strengthened? Should re-entry be regarded as a phase of the course, during which both the teacher and the Head need support? Might a conference for Heads at the time of re-entry, or earlier, encourage Heads to plan sensitively to support the teacher, and to make use of new knowledge and attitudes over a continuing period of growth? Colleagues, too, need to be involved in supporting the course-goer on return. As they showed their colleagues round the exhibition of their work on the course, some teachers were fearful of criticism, or perhaps of seeming to know too much. In a school with such an atmosphere, it may be necessary for a teacher to prove on return that no fundamental changes have happened. Teachers from other schools, however, found their colleagues interested and excited, and enjoyed sharing ideas with them. If a teacher's re-entry is not supported, much of the benefit of the course will be dissipated; it is very difficult for an established teacher to make any considerable changes in teaching style, so interest, encouragement and expectancy of change are necessary to help the teacher to grow.

CHAPTER 7

Case studies — other substantial commitments related to professional development

7.1 Introduction

In this chapter two events are discussed which were of a very different nature from any described earlier. What they have in common is that the participants in both events were required, at least temporarily, to employ a style of teaching that allowed pupil-initiated problem solving of a much more substantial nature than is usually found in primary classrooms. Moreover, in both cases, in order for the teacher to continue to participate, it was necessary to continue to use this teaching style, so that, unless the teacher dropped out, circumstances required continued classroom experiments in problem solving.

The first part of the chapter discusses INSET events related to the Open University course *Mathematics across the Curriculum* (The Open University 1980). The second event described was not intended as INSET at all: the teachers involved were asked to participate in a research project on strategies and procedures of mathematical problem solving, in which their pupils were required to engage in problem-solving activities. By allowing these activities to take place in their classrooms, the participating teachers found themselves teaching problem solving.

The major issue raised by these case studies is how far teachers on whom events force an innovative teaching style are likely to continue to practise this style, when they are released from the compulsion to use it.

7.2 An Open University course and related INSET

7.2.1 Structure and monitoring

The Open University provides two courses in mathematical education, which are intended for teachers; these are *Mathematics across the Curriculum* (The Open University 1980) and *Developing Mathematical Thinking* (The Open University 1982). Both are half-credit courses lasting for one year (February to October), and both involve a substantial component of classroom work, as well as personal study, the writing of assignments and an examination. Individual teachers may take these courses as part of an Open University degree programme, or they may take single courses as associate students. An LEA may also make a block booking for a group of its teachers to follow the course together. In this case, the LEA may expect the teachers to make some return by providing INSET for colleagues based on what they have gained from the course.

Both courses are suitable for non-mathematicians as well as for mathematics specialists, and they are taken by substantial numbers of primary teachers. It was possible to monitor parts of the *Mathematics across the Curriculum* course and related in-service work that had developed from it locally. Because of the length of an Open University course, however, it was not possible for the researcher to follow up by visiting teachers after the completion of the course. The following plan of monitoring was adopted. The researcher attended the four Saturday day-meetings of a local group who were following *Mathematics across the Curriculum*. These four meetings constituted the complete personal attendance required during the course, apart from the three-day Easter School, which was an essential part of the course and which the researcher was not able to attend.

The researcher also monitored two activities that arose from the *Mathematics across the Curriculum* course. A very important component of this course is 'real problem solving', in which a teacher, with the class, finds and solves a real problem which is worrying the children. Examples of problems that have been tackled are children's difficulties in the playground in a school in which infants and juniors have break at the same time, and producing a school T-shirt. The mathematics used in real problem solving is noted and considered. The Open University made the real problem solving component of *Mathematics across the Curriculum* available by itself as an INSET package for groups of teachers in particular LEAs, and the researcher attended an LEA group that was using this package. In another LEA, a block booking was made for a group of teachers to follow *Mathematics across the Curriculum*. The next year, they ran an INSET course for colleagues, and the researcher attended a session of this course.

Other substantial commitments

7.2.2 Mathematics across the Curriculum

In all OU courses, the resources from which participants are expected to learn are in the form of printed materials, TV and radio broadcasts and tapes. Written assignments are submitted at intervals throughout the course, and there is a final examination in November. Local group tutorials are arranged at intervals of a month or so, to give personal contact with tutors and an opportunity for discussion of the course materials.

The first Saturday tutorial for this course was held in March, the course having started in February. Seven teachers, three infant, three junior and one secondary, attended the local group meeting with their tutor. The group operated by informal discussion, loosely structured by the tutor. Some worries were expressed early in the meeting; some of the teachers were already concerned about the time the study of the course materials was taking, and there was some apprehension about the approach of the first assignment:

> I've always panicked about assignments — I never know what to do ...
> I have a fear of planning and writing ...

The tutor led the teachers to look at what they wanted the course to do for themselves, for the group and for the children. All the teachers said that they wanted to do the course not for the qualification but because they did not want to stagnate. They expressed the hope that it would make them more mathematically aware, and more confident. They wanted the same things for the children — more mathematical awareness and more confidence. One teacher wanted to analyse what she was doing, and hoped she would be better able to support her arguments to colleagues. The tutor saw the group meetings as providing encouragement and support — the teachers were in an innovative position when they carried out the real problem solving component of the course, but it would also be a position which might pose them threats. The group, however, were more interested at that time in seeking support by discussing the looming course assignment, which proceeded the real problem solving component; the remainder of the meeting was devoted to this.

By the second meeting, the Easter School had taken place, and the teachers were beginning to work towards their real problem solving project. The Easter School had been an unsettling experience for some of the teachers. They had themselves been made to undertake an experience of real problem solving in groups; in order to help the teachers to realise the effect of the group leader on the work, the group leaders had deliberately adopted an 'authoritarian' or a 'laissez-faire' approach to leadership, or a 'middle-of-the-road' approach. The course members would be expected to apply this experience during the rest of the course, and to think about the effect on their own children's work in real problem solving caused by the teacher's style of leadership.

At this second meeting of the group, the teachers discussed their experiences at the Easter School:

> We were being put through the process from the other side ... I was angry and frustrated ... It was easier for those who were in the laissez-faire situation ... The authoritarian one made us really mad — and even madder when one found it had all been set up ... I got so frustrated because the group could not understand me — I felt just like a child would.

> But I did feel I had learnt a lot about my role with a class.

One of the leaders who had played an authoritarian role at the Easter School was a teacher who had done the course some time earlier:

> That bloke I could not accept ... I was so angry — all I felt was 'He's only a teacher helping out at the OU' ... but I'm doing some in-service with some other teachers ... how do people react to me?

> It was better to have a teacher than the tutor ... at least it was put in our own terms — the tutor just gave us a lecture ...

> We thought it was going to be mind-bending on problems .... but it was mind-bending emotionally ...

> Those who come on the course are the converted .... I can think of many who need to ... Having said that, we are none the less still learning ... we are looking at techniques for doing the things ...

> Did the tutors sit on the fence when they said, 'Keep what you have AND change'?

The group tutor tried to answer this:

> There will be the test of time ... what will reside ... will it take over completely? — totally? — or disappear ...?

The teachers then turned in discussion to a conflict between the requirements of the course assignment, which consisted of a write-up of their real problem solving project, and what might actually happen when they embarked on real problem solving, especially for those who taught very young children.

> It's important that the children choose the problem ... this will make it real to them ... So if they did not choose it, should we imply that they did in the write-up?

> With young ones ... you must talk it through one-to-one ... I have sat and questioned them ... and I gradually got hold of their problem ... but you cannot use the work unless the children do choose ...

At the close of the session, the tutor reminded the teachers that there needed to be some evidence in the assignment that the children had

Other substantial commitments

used some mathematics as well as other skills when they tackled their real problem.

The next group tutorial meeting was not until September, and some of the teachers felt rather flat. They had finished their real problem solving projects before the summer holidays, and some of them had lost impetus over the holidays. The end-of-course examination became top priority in their thinking from now until the end of the course. The tutor tried to minimise its impact — not altogether successfully:

> They are saying, 'We are sorry we have to ask you to do an exam, but we've got to, and therefore we will do our best to ...' So I want the classroom situation to come through ... they will give you adequate scope to relate this course to your own teaching situation — this is what it is about. It is about classrooms ... it's about improving the teaching of maths ... I don't think it is necessarily about just giving answers to questions ...
>
> (Tutor)

But, on the whole, it had all been worth while; after a further gloomy discussion of the examination, one of the teachers looked back over the course and asked: 'Would you advise others to do this course?' The response from all the teachers was positive, and one was very enthusiastic:

> I think ... it was something I've been looking for, for some time, and now I've found it, so that's why I personally am enthusiastic. I had been looking at a lot of the logic materials there are around ... but I hadn't the sense to see that the stuff I really wanted was in real live situations ... I didn't have to go and find material ... it was all around ...

Finally, before the group dispersed, the teachers talked about spreading the message to others:

> It is ever so difficult being just one ... people might be interested, but they've got to be more than just interested to actually do the course themselves.

> One LEA did ... they had a group of teachers do it ... and then they fed part of it back to other teachers ... and there were official meetings ... and this is another way of spreading it — you do small bits of the course, but not for a qualification ...

> It is valuable for teachers to talk to other teachers .... to do the 'in-service' with other teachers ... but you need to have something to be able to give them ... and this has helped ...

7.2.3 *A real problem solving course*

The researcher visited a course in another area where a group of teachers were working on the real problem solving package, which the OU had extracted from the *Mathematics across the Curriculum* course to make an independent INSET package. This package contained

Other substantial commitments

materials describing for the teacher the idea of real problem solving, and suggestions for ways in which it might be implemented in the classroom. During the course session that the researcher attended, the teachers reported back in small groups on problems they had tackled with children. Some had come up against difficulties:

> It was on lunchtimes ... the size of the playground ... we began, but we didn't get very far ... we came up against the attitude of some older teachers not liking children probing into things in the school.
>
> ... this is why I chose something which involved my class only ... it did not encroach in any way ...

The tutor pushed the whole group towards personal reflection on attitude changes that had taken place in themselves and in the children.

> Perhaps as we get older the corners get rubbed off and we are ready for change ... and these ideas came at the right time for me ...
>
> I establish a working relationship more quickly with all the children ...
>
> ... for me it's not all of them ... a group has emerged ... I have respect for these, and their ability to make decisions .... but I don't like the way they have sat on others ...
>
> I have certainly appreciated learning from others on this course ...
>
> ... basically we seem to be on about a change of relationships ...

After coffee, the tutor tried to probe where the course had let people down. He led off by talking about his own fears:

> I was frightened about running the course in this way ... I felt a course should have some substance and it should come from me ... but the substance all came from the packages .... I felt out of control ... I kept wanting to go through the course books with you ...

The question why some teachers had dropped out was raised:

> We didn't know what we were taking on ...
>
> Perhaps they can't face this approach ...
>
> But the regular input was stimulating ... three meetings would have been too few ... I would just have put it away as an idea then and never used it ...
>
> We have had to think hard ... you don't assimilate unless you discuss ...
>
> I was cynical at first ... thought this is just the latest panacea ... I realised it was not a panacea after two sessions, when I began to read the materials ...
>
> I'm convinced after it all that this work is in the social context — it's not maths ...

Other substantial commitments

A month later, a full day end-of-course evaluation session was held. Several members of the OU team came, with the course members and the tutor. At first, they talked about what the children had learnt. The question of whether this was mathematics reared its head again.

> ... the mathematical skills were peripheral and the children all knew them before ...

> I don't see the problem solving as maths but as an organisational technique ...

> Is it because of the topics we chose? ... there might be more maths in other topics, but it could still be maths they knew ...

Then there were the political problems in school:

> The children have a lot of good ideas, but they don't expect to be able to do anything with them ... this is not normally what we do in education ... it was a shock for them to have to follow it through — and it was for me, too ...

> ... they have come up against political problems ... the caretaker and the dinner lady ... you can't upset adults ... you can say 'No' to children ... the teacher may not be able to face other staff at that moment ... How do you get changes? ... the adults want playtime together and the children don't ... and the adults win ...

After further discussion, a list was made of the mathematics that had been used in solving the problems:

matching one-one	using a grid to enlarge
using symbols	pie charts
graphs	percentages
counting, adding, tallying	estimation, checking
sorting and classifying	calculators
timing	costings
measuring length	area
scale drawing	recording and representation
money	coping at the cash-and-carry
spatial awareness	logical sequencing

> ... but what we don't know is how many skills were already there and how many had to be learnt ... or which were dormant because we don't use them ...

The tutor asked what the teachers had learnt about the children:

> ... some bright children were disoriented by not doing sums ... but some of my 'less able' maths people were asking the most incisive questions ...

> I've done something like this in language ... but I'm worried because I'm not monitoring individuals' progress ...

> ... much of the change in the children has been in attitude and they have NOT learnt any maths skills ...

Other substantial commitments

> I feel very much the same about it as the last speaker — I cannot see a link to the central core of mathematics ...
>
> ... we as a school are now thinking about skills in context, and we have abandoned our scheme — our aim is to stop using our content scheme for a time, so as to try to get down to the real fundamental ideas for our curriculum development ...

The topic for the afternoon discussion was what the teachers themselves had learnt from the real problem solving experience. This was difficult for them to talk about — for a long time they continued to talk about the children. Finally, they were asked to write down an experience that they had had, and what they had learnt from it. Then they went round the group and reported on what they had written.

> ... children using non-conventional methods of doing calculations ... multiply by 5 is double twice and add one ... on many occasions children were using maths much more creatively than I gave them credit for ... I became aware of their maths sense ...
>
> ... at the lunchtime club they wanted some dangerous activities ... what should I do? ... Did I forbid it? ... we left it in but discussed the dangers ... the democracy had to go on ... I feel it has built up a relaxed atmosphere for discussion .... and an acceptance of a changed teacher's role ...
>
> I learnt to let them make mistakes ... but I need to clear up in my own mind at which point one does need to step in ...
>
> I must not underestimate the thought level the children can achieve ... they were seven-year-olds and we had a problem which seemed to need percentage ... but they found a way round ...
>
> I allowed things to get out of hand — so that I could draw it to order later ... so I could point out their lack of time effectiveness ... I feel there is a lot of waste of time ... and I need to get this all thought out ...
>
> I was amazed at children's ability to cope without constant reference to the teacher ... on the scale drawing, they realised it was incorrect, but they found a way of adjusting it ... they had confidence to work without an adult and they worked well ... I must be far more trusting and not feel they *can't* ...

At tea-time, several course members said they felt that they had observed children more, and had thought more about questioning, through the work they had done in real problem solving.

In this INSET package, the written component — a compulsory assignment in *Mathematics across the Curriculum* — had been optional, and there was discussion about whether the written component should remain.

> I'm worried about the written work — how much value is it?
>
> I felt mine was worth doing because it clarified it all for me.
>
> If it is asked for, then it MUST be demanded — but should it be asked for?

Other substantial commitments

> If it had been demanded I would not have joined the course ... is there a necessary function for written work when there is no exam?
>
> ... there is a real place for reflection ... writing makes one order one's reflections ...
>
> We need to know at the beginning what is the written commitment ... also it would help to know how to present a report ...

Finally, one of the tutors pointed out that the real problem solving approach was one of many strands in mathematics teaching, and that part of the teacher's role lay in deciding when to use it. The group considered whether to stay together, and to arrange further meetings, but some of the teachers were going on to further work with the OU, and it seemed unlikely that this group would do more real problem solving — except as individuals.

7.2.4 An INSET course arising from *Mathematics across the Curriculum*

One LEA arranged for five of its teachers to take *Mathematics across the Curriculum* as a group; the next year, these teachers did some in-service 'give-back' by running a course for the LEA. One full Saturday was spent on measurement and precision, based on a topic which had been studied early in the OU course. Then — taking time to plan carefully — the group of teachers leaders arranged a weekend on real problem solving, modelled on the Easter School they had attended the year before. They used the three approaches to leadership they had experienced: authoritarian, laissez-faire and middle-of-the-road, and after the weekend course the members of the LEA course went home to do some real problem solving with their classes. After this, the participants met in local groups, and finally they came together for an evening meeting to review the progress made.

The group leader asked the teachers to discuss not what they thought should have happened, but what actually did happen.

> We were looking at PROCESS education — education for CAPABILITY ... to have skills to cope ... in particular, to be capable mathematicians ...
>
> So we looked at problem solving ... its affective side ... the emotions ... the frustration — especially when the purpose was unsure ... and also the sense of achievement when we have stuck together ... we need to understand the affective side ...
>
> Then we looked at the cognitive side ... where we need to define, pin down, examine, hypothesise, list ...
>
> Again we looked at the mathematics ... you remember the modelling diagram ... there is a need to catch the flavour of this process in schools — with the movement between formulation, analysis, validation, etc. ...

The teachers then reported in groups on what had happened in their classrooms when they attempted real problem solving:

> I told them I had been on a course, what had happened ... the enjoyment and the frustration I had felt ... so let's try some of this in school ... They were 8–11 year olds ... so let's explore any complaints we have — complaints — that's what their problems turned out to be ... for the first few days their ideas were rather wild ... it only calmed down because I gave a gap before we came back to it ... then in considering how feasible each idea was we gradually cut back ...

> I started immediately I got back ... I found they weren't used to talking .... they didn't easily see they had any general problems ... at first they only talked about individual problems ... it was hard to find any common element ... they didn't listen to each other ... There was an interesting problem of one boy who decided he resented the teacher not teaching — which he thought was the teacher's job ... we eventually finalised on boredom at lunchtime and what to do about it ... there was a certain resentment at having to discuss and do something different .. this was a real eye-opener to me ...

> ... everyone wanted to have their say ... until one child who had a quiet but persuasive manner came out with something that was puzzling her ... why there were birds in her garden at home — and she was obviously enjoying them — and yet there were so few in the school playground? ... this led on to an interest in this topic ... to find out why, and to encourage birds into the playground ...

The group responded enthusiastically to this account, and were very appreciative of the way in which such an uncontrived idea had developed. One member, however, had adopted a very directive approach:

> I went into my classroom and wrote RUBBISH on the board ... At first the children ignored it ... but eventually it led to a discussion of the word and its meaning — for instance, football teams you don't support ... later, having been encouraged to look locally, they saw litter ...

Later in the evening, the leader asked the group to write down, and then to share, what they thought had been the value of the real problem solving experience for the children, and for the teacher. Among the benefits the teachers thought the children had gained were:

> ... a realisation that they could work together as a group and get pleasure from it ...

> ... learning is fun and there is joy in sharing ...

> ... learning to listen and to accept other's ideas ...

> It built trust between the children and the teacher ... It made them realise they had to order activities— some things had to come first ...

Other substantial commitments

> ... it took them outside the classroom to consult ... and it showed the need to practise certain skills ...

> They learnt to work in groups — they were all very self-opinionated, and so they took turns to chair the group. They realised that graphs had some relevance other than just pleasing me as the teacher ...

> ... to listen, to co-operate and to order ideas ...

The group leader had noticed only three mathematical topics in the teachers' accounts — ordering, graphs and measuring, and called the group's attention to the emphasis they had placed on personal qualities.

> ... perhaps next time they will have developed more of these relationship skills ... and other things will emerge ... you cannot get good learning until you get good relationships ...

At last, the teachers started to talk about what they themselves had learnt:

> ... to be patient and to realise that the children can do it ... I must go more slowly ... children who cannot write can communicate verbally ... do we have children writing too much — should they write less?

> As a Guider, I let the girls solve problems — but I don't carry this attitude into school ... I now feel I have the confidence to do this ...

> I interfere too much ... the children can get on fine without me ...

> I needed to accept what they decided ....I had to stand by them ... I needed to be better prepared ... foreseeing them asking, 'Have you got this?' ...

> ... the children were shattered by other adults who rejected them ... what you can do in a school is very much defined by your environment — one is not an island ...

This led the teachers to talk about the other adults in the school, and about relationships with them:

> ... three of the staff are entrenched in their own classrooms ... I could not share ideas ... they would not want to know ... but eventually the other staff had to be involved, because the problem was about playtime ... the children wrote to them with their ideas ... the staff wrote back and did not think they were good ideas — in fact, they said they were bad ... It doesn't hurt children to realise that adults may have different views ... but it is very difficult to stay loyal when the arguments used are unreasonable and the children can see that they are ...

> How does one shake other people's gut reactions? ... It was done for us at the Easter School ... you can't tell people about it — they need to go through it ...

> ... ideally, we need a real problem solving course for Heads ...

> ... But, in my school, other classes do want to do it — the children, that

is, and the teachers don't know how to help ... the children are going round in circles without guidance ...

'So', the group leader drew the discussion together, 'everyone has a real problem — how to integrate in-service into the school'.

An approach to mathematics teaching as innovative as this — through finding the usefulness of mathematics in dealing with children's real problems (and those of the school) — cannot stay behind closed classroom doors. It inevitably moves out of its seclusion to affect the school as a whole. If the school, and the other teachers, are not supportive — or are openly hostile — the real problem solving approach will be changed by its environment. We may recall that, in the parable, the seed that fell upon stony ground sprang up, but withered and died for lack of nourishment.

7.2.5 Discussion

This Open University course is the only course monitored by the project that is an *award-bearing* course; teachers who successfully complete *Mathematics across the Curriculum* are awarded one half-credit, which they can use toward the completion of an OU degree. Thus, the courses are academic, in the sense that there are books and other course materials to be studied, assignments to be completed and, finally, the students take an examination to round off the year's work. Such a course is time-consuming and produces the usual anxieties in teachers who, since they left college, have become unaccustomed to writing essays and taking examinations. However, many teachers take OU courses, and the stress associated with these particular courses should not be greater than the stresses associated with any other OU course.

However, *Mathematics across the Curriculum* has a particular feature that sets it apart from every other Open University course taken by teachers. It requires the teacher to carry out a particular — and innovative — style of teaching, and it strongly suggests that this style is desirable and should be carried on as a regular part of the teacher's repertoire, and a normal part of her mathematics teaching. A teacher who was not able to make a reasonable shot at real problem solving during the course would presumably not be likely to be very successful on *Mathematics across the Curriculum*. A good deal of support is provided for the teacher, in the shape of regular tutorial group meetings, the intensive and social experience of the Easter School, the insistence on assignments, and the goal of successful completion of the course.

The style of real problem solving is not one which will be carried on successfully for very long by a teacher whose colleagues and Head do not understand and support what is happening, as we can see from the teachers' reports of their activities at group meetings. Not only does an individual classroom become a more democratic place, in which both the teacher and the children need to respect each other's problem-solving ideas, but the organisation and ethos of the whole school may be subjected to the children's scrutiny and criticism, and to their requests for changes. It is very desirable for children to use mathematics as a problem-solving tool in all their enterprises, and the particular enterprises undertaken in real problem solving are very likely to produce great involvement and motivation among the children. However, these activities cannot be confined to mathematics. They ensure that mathematics, and problem solving, will invade the 'real life' of the school.

The teachers' reports on their experience of real problem solving make us consider whether any type of innovation can flourish in an isolated classroom in a hostile or indifferent environment. How far is the continuing support of colleagues — and perhaps of interested people outside the school — necessary to keep the innovation alive, and to help it to grow and develop? Most innovations in the teaching style of mathematics involve changes in classroom relationships, even if the innovation is only to allow the children to move around freely during mathematics time, or to trust them to mark their own work, or to encourage them to put forward mathematical ideas for themselves, rather than always expecting the teacher to provide the mathematical input. Is it possible for changed mathematical relationships between teacher and children to flourish in a single classroom, if they do not flourish in the rest of the school? And what is the effect on a teacher of an innovation which was inaugurated, but which did not flourish? Sadly, the project ended before it would have been possible to go back and discuss the long-term effect of real problem solving with teachers who had done the *Mathematics across the Curriculum* course some time previously.

Another development from these Open University courses is the use, by LEAs, of teachers who have taken the courses to disseminate the ideas to others. When the researcher attended a meeting of a group who were following a real problem solving course, which was led by teachers who had done *Mathematics across the Curriculum* the year before, the atmosphere was very positive and dissemination was being undertaken successfully. It may well be that taking part in further dissemination encourges a disseminator to keep the innovation going, and to continue growing by exploring further the new teaching style and methods.

7.3 Informal INSET based on a project

7.3.1 The SPMPS project

The Strategies and Procedures of Mathematical Problem-solving (SPMPS) project (Burton, 1981), supported by the SSRC during a two-year period, tried to determine whether mathematical problem solving of a non-standard kind was feasible in the 9—13 age-range, whether children could acquire problem-solving strategies by being encouraged to solve problems, and whether there were any problem-solving skills that children of this age could learn and which would improve their ability to solve problems. During the project, a pack of sixty problems was developed, the problems were analysed, and they were then tried out in class by children. The project's research assistant visited the classrooms, and the teachers were also asked to report back on their children's experiences. The project ran for a term in each class — usually for one lesson per week. The children were given pre-tests and post-tests, and each school was asked to provide a control group, who took the tests but did not have any problem-solving experiences.

The project was not designed to have very much teacher input; at an initial briefing meeting, the teachers were asked to give the problems to the children and let them choose which problems to tackle, to encourage them but not to show them how to do the problems, and to observe what happened. A 'starter book' was provided, which teacher and children could work through together, solving the problem of 'How many squares are there on a chessboard?' as they worked through it. This booklet suggested a number of problem-solving strategies such as reading the problem very carefully 'Read it — REALLY read it', choosing helpful paper (squared, dotted, triangulated, etc.) to work on, and doing SOMETHING rather than staring at a blank piece of paper. Then the children, and their teachers, were on their own.

The project was a development from work done by a group of teacher educators; this group had included both the present writers. For the project's fieldwork, teachers and their classes were recruited by group members, and the group members visited the schools and provided the teachers with a limited amount of support. Although the project did not have an INSET focus, it was thought that it might have had some incidental effect on the professional development of the teachers involved, because they had been asked to work in a way that is not at all traditional in mathematics teaching in the 9—13 age range. The researcher therefore visited nine of the schools involved, and interviewed ten teachers who had taken part in the project, between one and two years after their term's participation in the project had ended. She asked them about their initial reactions, and the children's reaction, about whether they had shared their experiences with any of their colleagues, and whether there was now any residue in their mathematics teaching. Two of the schools visited were primary schools, while the others were 9—13 middle schools, which had some mathematics specialists on their staffs. In these nine schools, six of the teachers interviewed had undertaken the problem-solving project with 10—11 year old children, and four with 11—12 year olds. Four of the teachers were by training specialist mathematics teachers, two had become mathematicians by conversion, having initially trained for another subject, and the remaining two were generalist class teachers who were non-mathematicians.

7.3.2 Teachers' initial experiences

One of the specialist mathematics teachers was already teaching in a way that fitted in well with the organisation needed for the project:

> I had tried other problems before in odd lessons, so I was looking forward to it as a chance of making it a permanent part of the course ... We have four double maths lessons a week — two of them were on the SMP cards, one was on basic number work, and the fourth one I used for games and puzzles ... so to a certain extent it was an extension.

All the other teachers had great adjustments to make, and difficulties to overcome, although in some classes the children's attitudes helped them:

> I'm always on the lookout for new ways of tackling the subject. Particularly ways that appeal to the less able, and particularly ways that don't show a less able child against an able child ... There were great difficulties with it initially ... far too much material that we weren't familiar with. We didn't know the material well enough when we first started and, because of the amount there was, it would be quite difficult for anyone to, unless they had quite a lengthy period when they could have worked with the materials themselves ... now having worked with it for 18 months, and worked through these problems with different children, it's much easier now than it was. It was very, very difficult.

> It did strike me as being something so different from what I had come across before ... the thing that did strike me was the enthusiasm generated ... It was really refreshing to see it, and the children were really keen to get on with the next one, as opposed to going through ... wanting to do well and wanting to succeed ... It wasn't something that I was looking for ... it really hit me that the children were enthusiastic about it.

In every class there were some children who enjoyed the problems, but they were not universally well received:

> ... I wasn't sure how the children would take to it. I think in the end they did enjoy it. It was a challenge to the majority of them. There were some of them who found it an absolute chore ... they weren't always the less able ones. They were the children who never thought further than they needed to. They were really quite bright children, but they didn't push themselves, and they found that, to do this problem solving, they really had to think further than their little world. ... There were some children who wanted an answer, and if they couldn't get an answer this frustrated them ... they expect just to be told something, and they are not prepared to question and to think for themselves.

Other substantial commitments

In this particular primary school, there appeared to be some disjunction between the school's stated ethos and how it was implemented. The teacher described what the school's aims were, and then she responded to a question from the researcher about any mathematics her children had enjoyed doing lately.

> In some ways I think we've found it a bit surprising that some of them didn't respond better than they in fact did, because the whole ethos of the school right from infant level through ... is one of freedom in a mental approach, and investigation ... in the end you wonder if it is beneficial, has it paid off ... or are children just naturally like this ... and however much you encourage this sort of investigatory approach, they are still looking for an adult putting them right.

> I think my top group have enjoyed working on fractions at the moment, and we've been doing some cancelling work. We've been multiplying fractions and having to cancel, and they've enjoyed that ... just finding the numbers that cancel into each other. Looking and finding ... and anything associated with their tables ...

It was not surprising that this teacher rejected the project's ideas, and that problem solving did not take root in her thinking about mathematics teaching, although there was some evidence that it had helped her children:

> I was very impressed with the children who had done the problem solving because when they did their final test their logical steps were so much better than when they started ... but I felt it was an isolated incident and once it was finished it was finished ... and we never carried it on ... and the other teachers were in no way involved.

Some teachers saw enjoyment as all-important, and thought that children should not be forced to attempt problem solving unless they wanted to do so:

> What I do now is — we do it as a kind of a quiz thing ... some of them don't like it at all and don't really bother, and some of them love it and really enjoy solving them.

> I've tried to stress that they've got to enjoy it. I almost feel that if the children don't enjoy it, if children hate it, you are better taking them off it. Another reason why I want them to do problem solving is — I feel that children who don't like maths can enjoy this and it's not maths ... I wouldn't want to force it on the child ... If you don't enjoy it, there's no point in doing it ... that's the way I see it.

For other teachers, problems had something to offer every child, even those with very little mathematical ability. One teacher saw this approach as taking away the 'right or wrong' syndrome of mathematics from the less able, while the extensions suggested for some of the problems allowed scope for the more able.

> There was a chance here for a less able child to get success, virtually on

the same lines as an able, and this happened and it's happening now. John Peters was a classic case ... he had great difficulty with his maths ... Number is a new world to him, but he came up to me the other day and said 'I've done it — I've cracked it', and he was really excited, and I was excited for him. In fact, he had got it half right ... he was nearly there. I said a word to him and he got it, and that tickles me because he's got success in maths, and others can see that he's got success in maths. ... That's what I like.

The children, left without too much direction, adopted different forms of organisation:

> They didn't work individually unless they really wanted to, they worked in pairs or small groups, concentrating on the main thing of just starting off writing something down ... instead of just looking at a problem, saying 'I haven't got a clue how to solve that' ... We enjoyed doing it, and once we had been through the problems a few times ourselves, I used to find it much easier ...

> It was interesting that they started off wanting to work in twos or threes ... but by the end of the project a lot of the groups had split up because they couldn't agree on what was going to be the next problem ... or somebody got interested in something and wanted to go on ... so in fact by the end most of the groups had split up into individuals ... and I was quite surprised.

7.3.3 Language and communication

An unexpected spin-off for the project, and for some of the teachers, was in the area of language. The project team had realised that the first difficulty in a problem is in finding out what to do: 'Read it — REALLY read it';

> I think the actual reading ... what the problem is actually asking them ... they didn't really read carefully enough ... and I'm finding this happening now with the children I've got ... I don't think it does just come with solving ... they need — not to be actually taught how to read — but they need to be helped into reading a problem and finding out what is actually being asked of them ... It's not just maths ... in fact, problem solving wasn't just for maths — it was for all areas of work.

But there was much more to it than that. The problems engendered discussion among the children, and some of the teachers learnt much from that — even if they did not think the discussion was part of the mathematics.

> I think the language on this project far outweighs the actual maths, from what I've seen so far.

> I think the children enjoyed it very much, and they got much more out of it than maths ... and that was language development ... there was an awful lot of talking and discussion, and I thought that was its winning quality ...

Other substantial commitments

> I listened to the discussion and let them get on with it ... it was fascinating ... discussing with one another how they should approach it, and saying 'Well, that's rubbish' and 'We should be doing this' ... and just learning a lot about the children themselves ... let alone their mathematical ability ... I had a fascinating time listening ... incidental things that come out ... deciding who is going to do the recording — 'He's better at everything' — 'Yes, but his writing ... you can't understand that', which is lovely because that's what happens in life ... you've got to decide ...

In a pilot trial, the project team had found it impossible to understand from the children's writing how they had tackled the problems, because they had recorded their thinking so poorly. Consequently, the teachers were asked to stress recording, and to encourage the children to record their work for the project team to see:

> I hadn't expected the children to have as much recording problems as they did ... I thought that would come naturally, but it didn't ... it's better now, because we are conscious of it, and we try to encourage them to use different types of paper, if possible, because we've come to realise that one picture is worth a thousand words sometimes. Chris has got one or two of them to put some work on the walls, to display the results, and that has encouraged them to make a better job of it ... another purpose of doing that was to communicate to other people instead of just doing it to the teacher, who in fact knows the answer anyway.

The teachers found it difficult to help the children to develop their skills of recording what they had done; the work they usually did in mathematics did not seem to have taught them these skills. In one local group, the teachers held a discussion of the problem, and were divided on the desirability of insisting on recording:

> We felt the best thing would be that we took a problem and worked through it with the class and recorded, or said how we wanted the results recorded ... I suppose a lot of the teachers were in the same position ... we were not pushing the recording as much as the project wanted us to do. But it was obvious from the discussion that a lot of the teachers were not happy with the emphasis on recording ... they felt that the children were having a fair amount of difficulty working through the problems ... then at the end having to sit down and try to express what you've done ... by that time they'd solved it, and enthusiasm wasn't there any more. They were finding it very difficult to record ... It's very different from normal, because even with the work-cards they use, it's only a short step at a time and they're directed as to what they should do, in most cases. With this I think there's a big step to say 'Now write up the results, record what you've done'. That's why they find it difficult.

> Children are very reluctant to write down their resolutions ... some found it impossible ... and yet I thought ... that it was crucial if the child is going to derive any lasting impression from it ... it was very easy for some children who could do them ... some could be solved by trial and error, but it was important they realised that, and that there

Other substantial commitments

was some sort of logical approach to the trial-and-error ... they had to recognise that was what it was.

However, at least one child discovered, painfully, why it was necessary for him to record:

> One boy had worked out his problem — got his answer — and the bell went, and then he didn't meet problem solving for another week, and he said to me 'I've got the answer, but I can't remember how I did it'. And I said 'Why don't you know?' and then he realised he has to record every single thing he does.

Another teacher fought her way through to a new appreciation of the uses of symbolism in mathematics:

> We had done a lot of science work, and done lots of reporting ... and for some years they had done reporting ... I'm a bit sceptical about asking children to do it all the time in maths ... your main aim should be to get them to be confident and enthusiastic and feeling they can do it ... and perhaps you should vary your strategy ... I began to feel afterwards that they were then too verbose ... and really what benefited them in the end was that they began to see that there were briefer — more economical — ways of recording ... through symbols and diagrams ... I really had begun to realise this was happening.

One teacher, struggling to explain to the researcher the value of recording how a problem had been solved, and speaking before the publication of the Cockcroft Report (DES 1982), used almost the same words as the report about the aims of mathematics teaching 'an awareness of its power to communicate and explain ...' (§329) and 'mathematics provides a means of communication which is powerful, concise and unambiguous ...'(§3):

> One of its uses is ... it really brings home maths as a way of explaining things ... they've got their final solutions there so that everyone can see, and they see there's an actual importance in how they present it ... for communication.

7.3.4 *Developments in teaching*

The project necessarily produced a new style of classroom organisation, in which the children were much more free to follow their own ideas; they began to use the teacher as a consultant, rather than as a purveyor of right answers. This caused some of the teachers to think more closely about the role of the teacher.

> They have an attitude to you as teacher, and one thing that the problems do in some ways is it stops them coming up and saying 'Is this right?'. They come out and say 'Can I do it this way?' or 'Look what I've done', ... that gets a different response from me ... When they ask if it's right, you tend to look and say yes or no. It's really an enabling situation, isn't it, rather than ... sort of the authority figure — you know: 'I've got it all' ...

Other substantial commitments

One teacher reflected about the style of questioning he commonly used, and how his non-verbal communication to the children often encouraged them to play the game of 'Guess what I'm thinking', rather than to express their own ideas. In problem-solving sessions this was not possible, and he realised that questions could be taken in many ways.

> ... to be much more open-ended and ... sometimes the first answer can be the correct answer and that's it ... but sometimes it's important to hear — they've got a different answer — why? ... and what is their thinking behind it? ... and very often you can find that things are ambiguous ... and just because they have put the emphasis on something else ...

> Another thing I've found is that the question ... I load it so that I get the answer I want. The attitude you have with children tells them what you want, and if you want a positive answer your face tells them ... In problem solving, your role does change from being the teacher with all the answers to a learning situation alongside them ... I'm still learning, because every time they are involved with the problems I see something else.

One middle school teacher, who was not a specialist mathematician, joined the project after a mathematical colleague had taken part in the pilot stage. Her courage found its own reward, and her attitude as a teacher developed — she became more able to branch out and to explore mathematics with the children.

> I felt I would be absolutely out of my depth, which is quite interesting ... seeing Jane use them ... because she has the maths behind her ... her children would come up with things which I had never seen ... And doing it taught me a lot ... mainly that I was limiting the children to my knowledge ... and I think it is this fear ... I'm the teacher and I ought to know the answers ... And to be more true to myself in accepting that I didn't know what they were going to do. I just accepted it and I said to the children 'Look, we're just doing this together' ... and it did — it altered my role ... If they say 'How do you do this?' in maths, they expect you to know ... So in a way I changed my whole attitude to one of exploring ...

> ... it's very difficult to define what difference the problems have made to me ... all I can say is that I realise that in a lot of things, I limited them to what I thought the response should be ... I found that no longer was I saying to a child 'What is this?' and then realising he wasn't going to give me the right answer, and going on to the next child to get someone who can ...

One teacher reflected on the way in which schools encourage children to expect closed tasks and rewards for success, and another contrasted problem solving with the mathematics scheme that the school used, and a third commented on the conservatism of

Other substantial commitments

mathematics teaching:

> Something I don't feel my class do ... is the extension at the end. They solved the problem, they get the answer, their happy little faces when they get a tick and a comment ... and this is our heritage, as it were.
>
> The joy of SMP, although it has got its faults, is that they love it so much, they enjoy doing it so much, and I think it's because it's so well structured — the steps are so tiny that they can virtually all do it ... The essence of problem solving is that the problem is there and you've got to say 'If this doesn't work, try ...'. The suggested extensions are quite good as well. I enjoy it, I am going on doing it, but I don't do it as I did then ... I do it more as a game now ...
>
> When I first saw the problems I thought 'This is something that I'm looking for'. I've often thought that we tackle maths in the way we always have tackled it, and we never change the way we tackle it — we just do it that way because that's the way it's done ...

Some of the teachers did not recognise any great changes in their teaching styles; they were specialist mathematicians in middle schools, who had joined the project because they already claimed to be aware of the importance of problem solving and exploratory work in mathematics and thought that the project would give them further materials.

> Well ... I've made my questions more exploratory. You instinctively want children to — this is how they really learn — answering part of the question themselves. I wouldn't say it's influenced me a great deal ... just, I would probably use this problem-solving idea as a branch of teaching.
>
> It's difficult to know if it has actually changed what I would have done, or whether it has increased my awareness of the importance of problem solving ... but I have recently been doing some in-service work in school — helping the first-year teachers with capacity ... and, prior to having done the problem solving, I might have set up an experiment with pieces of Plasticine and displacement jars and so on, of fixed proportions, knowing what the answers would be ... but my approach this time has been more — there's a lump of Plasticine and here is all the apparatus you might find useful, plus some bits that were not necessary ... and set the problem ... like you do with children.

Some teachers valued the connections between the problems of the project and the problems of real life outside school.

> I can see this is setting them up for life ... life is problems and you've got to solve them ... it may be a money problem, and you've got bills coming in, and x amount of money coming in, and your first reaction is to panic because the bills are large ... but what you've got to do is to recognise you've got a problem, sit down and look at the facts and try to work out a strategy. So it's life. That's our philosophy in school, anyway ... we're trying to set them up to think ...
>
> Some of those problems were very good for their imaginations, you know.

> A couple of third-year boys ... not particularly brilliant ... we had some interesting links because they played snooker, and we tried ... of course the snooker balls were in a trinagle, and on a small table they only have four rows of balls, and on an adult table there's five, and they knew and they got the 10 and the 15 ... straight away ... and we got into triangular numbers ... It was the problems that had some bearing on their lives that really seemed to grip their imagination ...

There was some spin-off in other curriculum areas, as well as in mathematics, and the message of problem solving began to spread more widely, to other schools as well as those involved in the project. In one area, the theme of problem solving began to crop up at meetings of mathematics teachers across the whole area.

> I saw it widen maths at this school ... and it didn't stop at maths problem solving ... from there, for me, it went in other directions ... it crops up all over the place.

> Especially the 'What if ...?' bit ... That did spill over into my own teaching in all sorts of other areas ....

> We ... four or five schools have been given the task of looking at logic, communication and language in maths for the area as a whole, and three of us in that sub-group happen to be in this problem-solving group ... and it was agreed by the group that this would be an approach to language, logic, communications ... because problem solving in fact relies totally on those three things ...

> Something interesting happened this week — I went to a meeting of maths teachers on Tuesday, and problem solving was mentioned ... that problem solving ought to be a part of a mathematics syllabus — I've never heard that before ...

It is not clear how far the teachers who were involved in the project are continuing to use problem solving as part of their mathematics teaching. One teacher was delighted to acquire a pack of problem cards for her continued use, but others were evidently not using problems as much as they were at one time, and certainly not all the project teachers were entirely convinced in their own minds, although they were willing to experiment during the lifetime of the project.

> Originally when I was involved it was for a certain length of time and then it was over ... but then they said I could have the use of the pack of problems if I wanted so I carried on with them ... and it was only last term that I was re-involved again with them ... I have kept on with the problems with each class.

> I soon realised it was basically child-centred, very much so ... and therefore it's very demanding compared with normal class teaching ... Problem solving was additional to what I was doing, anyway, so I was trying to fit it in ... bearing in mind that it's not a part of the curriculum ... and at very much of an experimental stage ... Problem solving is problem solving ... the practicalities of what you can cope with given the time you can devote to it. If you did have a lot of free time perhaps

Other substantial commitments

> you could organise it a little bit better ... I do use it, but I use it in the summer term ... because of this problem about fitting it in to the curriculum, and the overcrowded groups ...

> ... the problem solving opened a new side of maths and it was a good side because the children obviously took to it ... but, unfortunately, when you look at the maths in the school today there's very little of it left ... I think one of the reasons was ... it was nicely boxed material ... and it was presented to the teachers and it wasn't very difficult for them to familiarise themselves with it ... But they also very heavily rely on the scheme we've got in the school ... SMP cards ... it's fairly structured and therefore they feel very heavily reliant on it ...

> ... it was difficult to persuade people that there was time ... there's no way of saying it's an easy progression, beause a child could spend three weeks at one problem ... It's convincing others, but at the same time you have a nagging doubt at the back of your mind ... that is ... at what point does it come back into mainstream maths?

7.3.5 Dissemination

The researcher asked the teachers whether they had introduced their colleagues to problem solving in mathematics. One enthusiast had successfully inducted a colleague into the project's work, but she was almost alone in her success.

> I've introduced Chris to it — she was the control group, so she wasn't involved at all. So I had to sort of put it to her ... what I tried to do ... I didn't tell her what I was told in the first place because I've learnt from experience ... and to me the important bit is that you have to give the child the right framework in which to tackle the problem — you've got to make sure that they are encouraged to use materials, to use different papers ... you've got to be very careful with your introduction and not to be there all the time telling them what to do ...

In general, even the teachers who were continuing to use problems themselves provided the researcher with a long list of reasons why their colleagues would not be willing to take part in anything so unorthodox. Some of the reasons were organisational and others were based on some teachers' limited understanding of mathematics:

> The other teachers that take maths aren't particularly ... they've got their own subjects they are more interested in ... so they are content just to get through the basics of maths, and they concentrate more on their own subjects ... If I could present them with the whole thing neatly wrapped, and say 'Here's the stuff — this is how you do it', one or two might do it ... I think with one or two of the teachers here, they would be a little bit concerned with children wandering about ... Children might possibly take advantage of the teacher ... the history teacher likes the children sitting down. I did do some Scottish work-cards with her class and she was very wary of them ... it's now part of her regular teaching, and she's quite happy with it ... This problem solving she was very wary of, and I'd have to ... tell her that movement and

Other substantial commitments

everything else she would have to accept ... So I'd have to make sure there was a reasonable amount of success to start with ... show her the problems, the most popular ones ... tell her the hints that would get the children started ...

... individualised work — we've talked about it, we had one or two discussions, but very few see it as the way ... maybe because their own knowledge is limited ... they want to stick very much to number work, and they can't see the need for maths to change ... it only comes when you study the subject in more depth and understand the changes that have gone on ... They envisage chaos and it's difficult ... I obviously believe in the philosophy of the problem solving, but there are times when you are not quite in control, or it's difficult to make sure that everybody's working as hard as they should be ...

There seems to be a minority of people ... like this *Sunday Times* crossword ... I think the majority of people are a little bit frightened of not being able to do it.

The teachers' own initial difficulties in organising for problem solving also probably discouraged many of their colleagues from wishing to get involved; anyway, it was only an experiment that would eventually go away:

... it was difficult spreading it to others ... we found it ourselves quite mind-bending ... and when one is taking a lesson and back in the staffroom you talk about things that have happened ... And you say 'My God, I think that was hard' ... and I think that might have put them off a little, especially as you are seen as the person with mathematics — if we had difficulty, what about them? ...

I feel it's something people haven't taken up because it's something that smacks of experimentation, guinea pigs, trial schemes, testing, something that perhaps won't be permanent ... and fair enough ... if they don't want to become involved at this stage, fair enough.

One teacher spoke pessimistically about the need to produce evidence that informal mathematics teaching and problem solving did work, if teachers were to be persuaded to change; then he suddenly realised that he did have the evidence, but that more was needed than evidence, if teachers were indeed to change their styles of teaching.

... unless you can show some sort of result — I don't necessarily mean written-on-the-page results — unless teachers can see something different in the child ... Maybe that's the only way of convincing laymen and sceptical teachers of the benefits of this approach ... Perhaps the way to show them is that you can gain facility by these methods. I'm beginning to wonder whether the old way is the only approach to facility — because my kids have tremendous facility with numbers, and it's not because I've taught them by the old way ... and if you can show that sort of thing ... you've gained a minor victory, but you haven't won a war — perhaps we never will, but maybe we ought to be realistic about that ...

7.3.6 Discussion

This case study raises the issue of the extreme difficulty, for a teacher, of moving to a new teaching style. The mathematical problems in the project's package were all non-standard, and did not use any particular pieces of mathematics; the children were expected to tackle them on the basis of such knowledge as they had. The project work demanded that the teacher allow collaborative work and discussion, and provide a range of practical materials and paper with different rulings — and allow the children to choose them as and when they thought best. The teacher also had to grapple with the difficulty that, although the children revealed a good deal of investigative imagination and problem-solving skill, they had very little idea of suitable ways of recording the work they had done.

For some teachers, the change was too difficult, and as soon as they decently could, they abandoned problem-solving work. Others fought their way through to some sort of resolution of their difficulties; perhaps being part of a project was helpful in making them persist and, in the end, for several of them, the new teaching style became easier than it had been at first. It is very necessary for any new venture to produce some success for the participants within a reasonable time, and the children's enthusiasm was an early success that was very sustaining for some teachers. When the children ran into serious difficulties, as they did about recording their work, the teachers no longer felt successful, and several questioned the value of insisting that the children recorded the resolutions of their problems. Recording was seen, perhaps wrongly, as a chore at the end, rather than as an ongoing support to the problem solver.

In their planning, the project team did not consider sufficiently carefully the teacher's role in problem-solving sessions in their classrooms. In fact, the design of the project involved work with children, rather than with teachers; the teachers were only there because the project team could not themselves work with enough children. The chief requirement for the teachers, therefore, was that they should not tell the children how to do the problems. But this role was so unusual and surprising for teachers that, in fact, it had a powerful influence on many of them — either a positive or a negative influence. Participation in the project was partly supportive of the teachers' new role; they had to struggle on, or to tell the project team that they were dropping out. But this was not nearly enough support; the teachers needed something much more than the project had envisaged, if they were to resolve the role conflicts that the needed style produced in them. In one or two schools, more than one teacher took part, and this was supportive. The visits of group members and of the project's research assistant were helpful, but much more would have been

Other substantial commitments

needed if the less aware teachers were to gain much of permanent value for themselves from the project.

However, for some teachers who persisted and became problem-solving enthusiasts, the project raised profound questions about the role of the teacher and about the values that children learn from their schooling. These teachers certainly seem to have developed their teaching style and the level of their thinking, because of the project. Perhaps they were already well along the road — it is significant that several of them were already following, or had just begun to follow, substantial INSET courses. Some were doing an in-service BEd, and others an MA in education, at the local university.

The experience of this project, unsuccessful as it was for some teachers, does give some pointers. A project can produce changes in teaching styles — at least for a time — among those who volunteer to take part in it. If the materials are sufficiently centred on the children's explorations, through problem solving or investigation, then the teachers must allow exploration, or drop out of the project. Built-in local support groups, in which the teachers could freely discuss their doubts and fears, and their difficulties in adjusting to the new role, might enable teachers to come more easily to terms with a less directive teaching style. Early success, and some sustained progress, are very important. A more positive planned role for the teachers would have been beneficial. They could have been involved in using their experience to package the project materials into an easier form for others to use. They could have been involved in indicating good problems to start on, in compiling lists of useful materials, and in disseminating the project among their colleagues, as well as in contributing to solving knotty problems such as that of recording. Guidance was also needed in how to integrate problem solving into an ongoing mathematics curriculum. None of these tasks was among the aims of the SPMPS project, but it becomes possible to see how a research project such as this might become a powerful INSET influence on those who take part in it.

CHAPTER 8

Interviews with teachers and INSET providers

8.1 Introduction

The work of this project has concentrated on *courses* of in-service education in primary mathematics. However, courses form only a part of a complex network of INSET which can support teachers in their professional development. Another important aspect of the INSET network is work within the school, in staff meetings focusing on the curriculum, in informal discussions with colleagues, and in the teacher's own reflection on the day-to-day problems of teaching mathematics, and in her reading and preparation; these also give learning experiences through which a primary teacher can develop as a teacher. In some LEAs, INSET within the school is seen as such an important contributor to professional development that the LEA provides teams of Advisory Teachers in mathematics; the work of these teachers is to visit primary schools and work alongside teachers in their classrooms on mathematics, providing support and leading by the example of their own teaching. Advisory teachers also often lead staffroom discussions, bringing the whole staff of a school together to work on mathematics.

During the project, the researcher had many discussions with class teachers, with Heads, and with INSET providers such as Advisers, Advisory Teachers and college lecturers. During these discussions, they often reflected on the roles which INSET of different types can play in professional development. These discussions, many of which were recorded and transcribed, form the subject matter of this chapter. They focused on several recurring themes, to which the speakers continually returned; these themes are used to organise the sections of the chapter.

Other interviews with teachers

8.2 Working with teachers in the classroom

Teachers do not always realise that mathematics can be taught in different ways from the method that they themselves use; nor do they always find it easy to develop a new style of organisation and teaching. Support from an Advisory Teacher or other visitor who works alongside them in the classroom can be a potent force for change. Here, an Advisory Teacher talks about how she tries to work in schools:

> I think one of the main objectives of my particular contact with schools is really to reassure and to give a bit of praise and encouragement ... and always work from the positive, of course. Sometimes you have to dig deep to find something that you can encourage and praise ... but that doesn't happen very often.
>
> The best way to encourage a teacher may well be to go in and work alongside that teacher. I don't find that easy .. I want to work alongside the teacher, but there again you are quickly trying to adopt someone else's mode of organisation — classroom organisation, as well as organising the maths programme ... and I find that very difficult ...

Advisory Teachers and other consultants need to be extremely sensitive to relationships within the schools that they visit; a relationship of trust with the teachers in whose classrooms they work is most important, if those teachers are to develop and change their ways of working.

> The staff were a bit defensive ... the first term was at the level of becoming trusted ... of becoming credible ...
>
> ... the first thing I find I have to do is to prove that I can teach ... so I go in and teach ... I am basically going in once a week ... I teach the class and I leave material behind ... when I arrive the next week I know where I'm picking up ...
>
> If I was teaching my own class I would not do it the way I'm doing it in there ... it's for a fixed period and I'm having to teach in a certain way ...

Working alongside other teachers makes Advisory Teachers acutely conscious of the most common problems and difficulties that teachers find in teaching primary mathematics:

> Listening to what teachers have asked me, there have been very common threads ... such as 'I know the children develop through experiences and practical activities ... but I had forgotten what the value is of this particular situation' ... This is what the teachers are saying ...
>
> They don't always put the emphasis on getting down with the children and enjoying the activity and perhaps learning alongside ...
>
> ... teachers seem to want to take away the crutch before the child rejects it ... I think it is because they feel they are not teaching properly

Other interviews with teachers

> .. they are not teaching fast enough ... You get the teacher who will say 'Yes, but what will the parents say?' ... or in the infant school, 'Yes, but the juniors want ...' ... or 'the middle school wants' ... or the high school ... These mystical expectations there seem to be ...

Schools and teachers are often anxious to be 'doing it right' in mathematics, and in their anxiety they may try to show the visitor what they think she will want to see:

> ... At first, it was just sort of showing me round ... then the head would invite the teachers into the staffroom and the most common question was 'Well, what do you think of the maths?' ... It was quite a shock when that question was asked for the first time ...

> And, of course, they are all doing mathematics at the level they expect you want to see .. the work-cards, the work-book, the written sums, the expectation of what you want to see ...

Some teachers find it very difficult to take a lead from a visitor in their classrooms, even one who comes from the LEA, and whose approach to mathematics therefore receives official support:

> In one particular school the teacher rarely follows up what I do, so what I do with those children is isolated ... and with infants you can't do something once a week and expect them to carry over and get anything from it ... in fact, that was a school where I very carefully spelled out at a staff meeting six months ago that my role was for the teacher, and I would expect things to be followed up ... but once you've said it you've said it ... you can't twist people's arms ...

> The teacher was very wary of me and she's not happy to work with me and therefore she has me working with her individual children and I do anything she says ... I produce every week follow-up and a report for the school on what I have done ...

> I thought I could use the mathematics as a means of getting in, to alter her classroom organisation ... I have been totally unsuccessful in this ... she's a very strong-willed woman, who keeps the class under control because of this strong personality ... so she in fact doesn't see the need to alter the structure ...

The next speaker is an Adviser, talking about the work of Advisory Teachers in his LEA. The pace of change everywhere is slow, and many teachers need continued support over a long period:

> An Advisory Teacher spent one afternoon for a term working in the classroom with some teachers ... on returning to the school a term later she found that no initiative had been taken in the meantime ... there seems to be a real need for continuing input ... and schools don't really seem to begin to cope on their own before about three years ...

Another person who is often in a position to support teachers by working alongside them in the classroom is the Head. Some Heads who have a special interest in mathematics make this a feature of their

Other interviews with teachers

staff development technique. One such Head told the researcher about her policy:

> On my way around the school, if I find that a teacher has got something wrong, or the language is wrong ... it's there and then you need to do it ... so I usually say to the children 'Now listen, children, I am working with this particular group, and Miss So-and-so is watching to see that they do it right' ... And then I'll show the children ... she suddenly realises that the wording is wrong, and she just didn't know how to do it ...
>
> I think you've got to help them through the children ... that you've got to say to the person 'I'm coming along to your room this morning — plan your day as you normally would and I'll give you a hand' ... and then if I do see any mistakes I'll do it with the children ... The usual reaction is 'Oh, I didn't realise that' ... Now they are not being criticised by me, no, this is the whole point ... Once they've lost their confidence, it's hopeless ... and you see we never say to a child 'You're wrong', so similarly with a teacher I would never say 'That's wrong'. It's a case of letting them 'think it out' ...

In some schools, it is possible that the mathematics co-ordinator may be released to work alongside colleagues in their classrooms. At the time of the project, this was a rare occurrence, and the researcher did not meet any co-ordinators who were working in this way.

8.3 School-based INSET provided by a college

In the next group of extracts, a college of education lecturer with considerable experience of school-based INSET in primary mathematics compared the effectiveness of courses and school-based INSET.

> It's got more value than teachers coming on a course ... they actually see it happening with the children ... I'm sure you've heard 'Yes, that all sounds very nice ... but it wouldn't work with my class.' ... But if you are in the classroom, it is their class and although they've got three children who can't read, you are coping with them ... they can see that the problems can be solved ...
>
> Group work ... 36 children — you can have, say, nine groups of four and you can have some measuring and some doing this or that ... and this sounds very nice in a lecture room, but then they go away and think 'Yes, but ...' Whereas if it happens in the classroom they can see it is possible ... this is the great benefit ...
>
> In a course, you get something up in a lecture ... and you give examples ... but then it's really quite difficult for the teachers to take it back and apply it there and then ... because it doesn't fit in with what they happen to be doing in their class at the time ... it is a difficult problem to take it back, even if one is trying to ...

However, even when the visitor sets up a different type of organisation and demonstrates that it can work, individual teachers need to be personally convinced that a changed style will improve their teaching if they are to develop in the ways suggested.

> It was a junior school ... the Head initiated it ... there were three staff in the lower part of the school ... one of them was very suspicious ... and another of them was a young girl who was hoping to start a family and she wasn't bothered either way ... and the other was very enthusiastic ...
>
> I did area with one class and weight with the other two ... and what happened was that I prepared all the work and took in all the materials ... and the teacher stayed in with me and worked with me, and afterwards we got together in the lunch hour and discussed what had been done ... That's looking at it with rose-coloured spectacles in a way, because teachers have other things to do in the lunch hour, but that is in fact what happened to a limited extent ...
>
> There was another problem — there was very little apparatus for weighing and so on ... they were into art really ... the Head is very inclined to art ... you know Heads spend their money on what their interest is ... so I shouldn't think it led to them getting more maths apparatus ...

Even though the Head had initiated the lecturer's visits, it would seem that only in one of the three classes was there any likelihood that the teacher would change, and that change might well die away through lack of support and through lack of availability of the needed equipment.

In another school, the Head's support for mathematics, and her leadership of her staff, made the lecturer feel that his work was more likely to bear some fruit.

> Another school ... the Head said her syllabus was just being typed, and could I look through it? I made one or two suggestions, and she said would I come back and discuss it with them ... They were the opposite of the other school — they were extremely enthusiastic. The Head was very keen, with a staff around her who were very keen ... The staff were used to staying until about 5.30 in the evening, and this makes a difference ... in the previous school, they weren't.
>
> ... junior and infant ... the syllabus was only junior ... We took the syllabus and went through it with a fine tooth comb and discussed it. If I said 'This ought to come after that ...' or 'This isn't correct, you can't say this ...' we would discuss it ...
>
> We got this syllabus sorted out to everybody's satisfaction ... and then the Head said the infant teachers, who had been in on our discussions, would like you to go through their infant syllabus, which they now think is a bit old-hat ... She gave me a copy and it really was pretty awful, so I had to be rather tactful ... We then went through the infant syllabus, tying it in with the juniors ... it was rewarding, because they were keen ...

School-based INSET such as this is extremely expensive in terms of the time of the lecturer, who can only reach a smaller group of teachers than if he spent the same time in running courses. The time spent is justified if the work is more effective than courses would be, but the experience in the first school suggests that such work needs to be prepared for in the school and followed up afterwards, and that classroom demonstrations of different ways of working may not be effective without such support.

8.4 The effects of courses

For most primary teachers, their chief experience of INSET is 'going on a course', rather than receiving a visiting Advisory Teacher, or other consultant, either in their own classroom or to take part in staffroom discussion. Here, a group of teachers reflect upon their experience of course-going, and on how far they manage to bring the message of a course back into their own classrooms.

> These courses that bombard you with all sorts of fabulous information which is bang up-to-date and everything else ... and the people that take the course are really enthusiastic ... But that science course ... I'm not very interested in science ... I went along just to see how it was presented and so on ... I got a lot of ideas from it, but I think you've got to have the interest yourself in that particular subject to bring it back to school. I still don't do a lot of science ...

> On the music course it was much the same ... They produced two schemes for us to look at, which both seemed absolutely ideal ... but then you went away and thought about it ... the amount of money it would cost ... and you were probably doing just as well with what you had ...

> Our facilities are very limited ... to go on a course where you use particularly good apparatus or whatever ... and you haven't got those ... well, then, that course isn't going to be particularly useful for you ...

> I went on a course on using Fletcher in the classroom ... it would have been helpful with a small group of children ... if I had an assistant in the classroom I could take a small group of children — it would be marvellous ... but not in the whole class situation ...

> People have said to me that they wouldn't dream of ever using Dienes' apparatus because of an experience they've found to be very unhappy ... either in a college of education for a couple of hours ... or on a DES course for a couple of hours ... or even in a local Teachers' Centre for a couple of hours ...

Course providers, too, are often very much aware of the difficulty of persuading teachers to take ideas from courses back to their classrooms, and especially of the difficulty of persuading them to change their minds about their style of mathematics teaching.

There's a sort of feeling around that courses are no good — they don't have any effects ... I don't believe that ... but they're not the sort of effect that comes from giving a lecture to a group of people and then they can write up notes about it and answer an exam question ... it's not that sort of effectiveness ...

One wonders how and what they take back ... our course was saying there is a logical development through number work ... it was saying that most people go too quickly and revert to traditional ways of teaching number, and that we don't agree with this ... so it was perhaps reinforcing some people who felt this but who were happy to have it said again ... It certainly made a number of people realise the conflict between what they were doing and what somebody else was saying ... I don't know whether they resolved the conflict or what they did about it ... but at least it produced a conflict ...

It was part lecture, part workshop ... and there'd be activities in which we'd work with groups ... and I think it was successful in the sense that people enjoyed them ... but of course I've no idea if they were effective ... I have great reservations about how much goes back into the classroom ...

I was brought up in an anti-tips-for-teachers age, but I think now that teachers want something fairly definite ... if they have one or two activities that they can go away and do with the children, it seems to me that they've got something out of it ... it might just make them go on and do something else ... I have a theory that if you can get one good idea from any course you're doing very well ...

I think you can run a course with one intention and the effectiveness is in quite other directions ... The individuals come from their situation, they listen, they probably hold on to things you didn't know you were emphasising ... so I don't think you can say that the aim of this course is ... and that is what I am going to look for ... the effect ...

In a day course ... you can't change people's attitudes or they can't change their own attitudes ... but a week's residential course can have a dramatic effect ... In that week you get very hard discussions and conversations and conflicts coming up, and there's something very intense about the whole business ... The really strong people I see in primary schools have most of them been on one of these courses ... I don't mean that the course made them strong, but it provided a focus for them ...

The only follow-up I've done was for the one-term full-time courses ... we used to send out a questionnaire, which said things like 'How did you arrange the room before the course, and how do you arrange it now? What books did you use before and what do you use now?' ... And if they told us the truth, and if we were asking the right questions, that course did have a major effect on the way most of those teachers worked ...

Thus, course providers tend to distinguish between the purposes of courses of different lengths; short courss can give 'instant input' or can raise issues that may need to be resolved later, while long and

Other interviews with teachers

intensive courses are necessary if attitudes are to change, and to give teachers time to rethink their mathematics teaching.

8.5 Teachers' needs at different stages

Teachers at different stages of their careers have different INSET needs, and may get different things out of the same courses. When they look back they are often aware of these different needs, and can identify the development they have gone through. A group of teachers discussed with the researcher courses of different types, and what they had gained from them.

> In the first few years . . . some courses were just a repetition of college . . . It was still fresh in our minds, anyway . . . we went along to several reading courses and that was entirely repetition . . .
>
> . . . perhaps for a few years you don't want new ideas . . .
>
> Occasionally you'd meet a reading scheme which was new to you, and you'd have look at that . . . That was about the only thing I did because it was all so fresh in my mind from college. . .
>
> I think the longer you're out of college the more you're aware that you do need to be rejuvenated and go through the whole process again . . . When young people used to come out of college they used to bring new ideas with them to other teachers, but we don't get that now . . .
>
> I am doing a Dip Ed . . . I find it very far removed from a class teaching situation . . . I find having been out of college a few years I miss the academic exercise . . . I enjoy that part of it and I think if I were going back that's what I would want to do. . . .
>
> It would be a secondment . . . a time for deepening or standing back . . . but that would only come after a few years . . .
>
> I was on a long course . . . I started on it for promotion . . . I needed something more than the Certificate, but once I started I realised that I got value from it . . . and it helped in teaching . . . It gives you more confidence in your decisions, because you can justify it more . . . because you know you can go to the right theory . . . it makes you more critical of yourself as well . . .

An Adviser discussed how he responded to these needs by trying to arrange courses in such a way that teachers who wanted direct classroom help, and those who had reached the stage of needing to reflect and justify their work, could both gain something from the same course.

> You might call it Level 1 and Level 2 in-service . . . when we have the mental arithmetic session next time, my intention is that it should be on Level 2 . . . it should be involving people in thinking what the purpose of working with numbers is . . . but the format of it is Level 1 in terms of things they can go straight back into their classrooms and do . . . it

Other interviews with teachers

seems to me that what I want to be searching for all the time is that sort of thing which is on both levels . . .

A common complaint about courses . . . was that it was great while we were there, but once we were back in the classroom we couldn't really see what we were going to do . . . I think I've got to bridge that gap . . . I haven't just got to pander to what to do in a classroom, but I've got to be able to defend anything that I say on Level 2 in terms of what it implies for Level 1 . . .

Some teachers work their way through the process of reflection on their own and their colleagues' practice, and take on an in-service role outside their own schools. A Head who had become a leader in mathematics talked about how she had reached her present position, and the support that teachers in a leading position could give one another:

I'll talk about how I got into it . . . I was worried about maths teaching because the children couldn't understand . . . and so I became quite involved in it as I went from school to school . . . . . . and because of one's own enthusiasm, people catch this . . . and when you open your mouth at a meeting or a course . . . people catch the enthusiasm . . . and I became caught up in talking at courses . . . not just about the activities . . . but perhaps more in a philosophical kind of way . . . and we started a maths forum, and I did a term's course . . .

Then you find you're asked to run courses . . . so I've run courses on infant maths, for which I am not trained . . . I am a trained junior teacher . . . and I was asked to talk about things like trends in infant number . . .

People ring, and if they want to see perhaps multi-base working in a school . . . or logic . . . the Adviser will direct them to this school . . . because there aren't too many schools where there is this sort of continuity . . . and they can see the whole thing from 5 to 11 . . .

And I think there will always be . . . among people that are interested in the teaching of mathematics . . . these discussion groups . . . and of course one is furthering one's own mathematics at the same time. Those of us who were taught as I was in the rote kind of way . . . and you start having more insight into it as soon as you start teaching . . . but I still am having insight into it, and analysing how I work things out . . . because that's all-important in helping children, of course . . . but also talking to other people about it and helping them to look at it . . .

8.6 Needs of co-ordinators

The last speaker was a Head who acted as mathematics co-ordinator within her own school, and who had reached a high level of professional development. Most mathematics co-ordinators are not Heads, but if they are to have real influence in their schools, they need to reach a similar point of professional thinking. However, even keen

Other interviews with teachers

co-ordinators, such as the next speaker, start at a basic level, and often the first jobs that need doing are equally basic.

> When I first got responsibility for maths I was moving to a new job. All I found when I got to the school was a maths cupboard ... that was it ... So the first thing I did was to take a stock check to see what we had got, and then I started signing on for courses to get myself 'genned up' ... anything to do with maths ... Then I got on to a working party on junior-secondary liaison; this meant I met other people and got support from outside the school ...
>
> Then you get to the stage where you are given responsibility for ordering stock ... you can buy what you want ...
>
> I changed school again ... there was a maths cupboard ... It was an absolute tip, so the first thing was to have a good sort out ... I decided to keep all the stuff more or less in the stock room, but tidy it all up and have everything put into boxes ... We had people using equipment again ... often they are afraid of using it because it's too hard to get hold of ... I didn't really have to help many teachers use the equipment, but when new things came they were shown to people at the staff meeting, and we talked about them ... and then they knew they were on offer in the cupboard ...
>
> And then we got to the situation — which was ideal — where the infant teachers were taking things out of the stock cupboard and keeping them. Other teachers grumbled, because they wanted it ... so now they were fighting to use it, as opposed to people not doing anything and the equipment lying in the cupboard ...
>
> In this school I'm starting all over again ... we have some spare classrooms and I want to set up a maths room ... Eventually I would like to have every classroom with basic material ... cubes, a balance ... and they would also know they could borrow things ... bigger pieces of apparatus would be shared between a year group ... That's the first thing I want to do ...

This co-ordinator was successful in ensuring that colleagues knew what apparatus was available, and that they made use of it. He had a clear division in his mind, however, between the co-ordinator's role and the Head's role in promoting the mathematics teaching in the school.

> We have got a new scheme to watch over — now I would see that as the Head's job, not the co-ordinator's ... The co-ordinator is there to make sure people have got things ... he is like the foreman, gets all the dirty jobs — he makes sure the stock is ordered ... he may have unofficial chats with people, but if teachers are in difficulty, I would see them turning to the Head ...

As he talked to the researcher, this co-ordinator groped his way towards new ideas about the two roles. In his mind, at present, helping weak teachers or teachers who disliked mathematics was not part of the co-ordinator's role.

Other interviews with teachers

> But ... but ... it's often easier to pluck up courage to have a word with one of your colleagues ... This is where the interplay between Head and Scale posts comes in ... yes, if a co-ordinator was really functioning this way, then if he found several people asking about a particular topic he would feed this back to the Head, and something could be arranged for the whole staff ...
>
> The other thing you've got to bear in mind is that you will always have teachers who are frightened by any scheme ... However, hopefully even the ones that don't like maths will use the scheme ... So a Head has got to make sure they are doing everything ... that is not the job of the co-ordinator ...
>
> No scheme will ever replace a teacher, but a good scheme may improve a bad teacher ... the only one who can really go into their class at first is the Head ... This is not the job of the co-ordinator, but he and the Head do work together and he may be able to bring up unofficial conversations in the staffroom about topics of concern ...

This co-ordinator still had a long way to go before he would see himself as being able to lead staffroom discussion or to work alongside colleagues in their classrooms. Perhaps discussion with other co-ordinators, or a longer course, might help him to move from seeing the co-ordinator as a provider of stock to seeing the co-ordinator as responsible, under the Head's guidance, for leading the development of the mathematics teaching in the school.

8.7 A teacher on professional development

Finally in this chapter, a teacher whom the researcher had come to know well discussed her own professional development:

> Since I have come to this LEA I seem to have been on some marvellous courses ... I'm not sure whether my last LEA wasn't really organised when I left ... or whether I wasn't ready to go on courses ...

What were you looking for in your last LEA?

> First of all, a maths course, because I was in charge of maths ... but ... we didn't seem to be a school that ... here, the schools seem to have the teachers' manual with all the courses in it ... and it's there, and it seems to be available ...
>
> And whether it was there in my last LEA ... we were such a large school that I was never aware of it ... we had 140 in the infants ... which is the same size as this school, just for the infants ... and in the infant department we did team teaching and we worked very hard and closely together ... we didn't have any energy left for courses, I don't think ...

As a new teacher who had entered the profession after a mature student's course, consolidating her teaching skills in order to make a full contribution to the team-teaching situation took all her energies:

Other interviews with teachers

> We did help each other a great deal ... we did a lot of supporting of ourselves ...

So, even if there were courses available, she did not feel the need to seek them out.

> *You did say — perhaps I wasn't ready for that sort of thing?*
>
> No — I don't think I was ... because we were so busy with the team teaching ... and that was more than adequate support ... It was a marvellous situation ... we ended up with four teachers of about the same age ... so that eventually we got on so well together ...

The researcher probed the reasons why the teacher had started to ask new types of questions about her teaching; it proved to be a combination of a growing dissatisfaction with some limitations in her own mathematics teaching, and the need for further qualifications to enable her to obtain a full-time post:

> I came back into teaching ... I was missing for a time ... and I was very dissatisfied ... first of all with the limited type of 'sorting' we were doing ... and so I went on to the logic ... I was very dissatisfied with myself ...
>
> ... and I had been on a two-terms' maternity supply down the road ... and while I was there I saw the maths diploma advertised ... and I thought first of all 'I want to do it' ... and secondly ... if I am going to go on doing just two or three terms here and there, I need further qualifications ... and that definitely entered into my calculations ... but the teaching situation there was such that we didn't have to do further things ... I had to be realistic about my future and go on and do other things — and that was what I wanted to do, anyway ... so the two reasons combined ...

She gained greatly in confidence from taking a very substantial course, even though it was a struggle at times, both personally and in bringing tensions with colleagues:

> I have gained so much more confidence since I went on the maths diploma ... it has given me such a depth to look at the way I am teaching ... I never was frightened about maths, but now it comes into everything we do in my class ...
>
> I felt at times 'Oh, God, I am not going to cope any more' ... you know — perhaps because we were going too fast, but then you had to fit in with the timing of the lectures ...
>
> I think it has left me with a feeling that — all right, I will pick up a maths book .... and I will look at what I can cope with ...
>
> ... this maths diploma — first of all, it produced tension in the school ... I asked if I could test certain children on certain things ... the first question they asked was 'They won't know which school it is?' — 'Don't put the name of the school on it' ... so having got over that hurdle ... we discovered that there were in the school quite a few children who

> didn't have concepts that we thought they had ... they then accepted that I wasn't really giving the school a bad name ... which I think was a fear ... and the subject itself creates tension, anyway ...

Finally, the teacher looked back on the reason — or rationalised the reason — that she had not gone on a diploma course earlier. She was probably very fortunate that circumstances forced her to take shorter courses first, rather than a demanding course that she might not have been able to cope with:

> I had tried to get on a maths diploma when I was in my last LEA, but they wouldn't accept me ... they didn't think I was suitable ... they were really looking for people who were teaching, say, nine-year-olds upwards ... for the longest courses ...

With the diploma behind her, she was co-ordinating the mathematics in her school, and leading an active professional life; she was beginning to be invited to take part in in-service work in her LEA, although she was still nervous about working with other teachers.

> ... what really started me off when I came here ... they had a course about young children learning ... and we went to different schools — and they were experienced teachers — and we made great friends with each other ... and from that another offshoot has happened ... the Young Children Association ... we have speakers ... and I'm on the committee of that ...

> ... going back to that last maths course in the LEA we were talking about ... I did talk to the Adviser about the next one ... and he would rather like me to be there again to help, so I would meet another set of teachers ...

> I would only take on something I thought I could cope with, because it is a bit intimidating ... all these experienced teachers sitting around you ... and I was a little bit like that at the last meeting, until we got down to it ...

8.8 Discussion

These teachers, from all their different points of view, demonstrate the range of support that is needed for full professional development — the short courses to give fresh classroom ideas, frequent supportive visits from the Head in the classroom, curriculum discussions in the staffroom, provision of a good supply of apparatus and resources, Advisory Teachers to demonstrate a variety of ways of working, longer courses to give opportunity for reflection, discussion groups in which mathematical ideas can be analysed, groups where co-ordinators can gain strength through pooling their experience, the opportunity for some to become INSET providers and to share knowledge gained. Very few teachers find it possible to take part in

Other interviews with teachers

this full range of INSET in mathematics; of course, many teachers who are not themselves oriented towards mathematics find much of their professional development through other curriculum areas, but still need some support in their mathematics teaching. However, the conversation of the groups of teachers quoted in this chapter shows the need and the value of variety in INSET provision in primary mathematics, and the importance of making a range of different types of opportunity available to teachers at different stages of their careers.

CHAPTER 9

The individual teacher and INSET

9.1 The model of professional development

In this chapter, some detail of the model of a teacher's professional development is filled out; this model was outlined in §1.6. There it was suggested that teachers seem to pass through four main stages of professional development, although these are not rigidly separated from one another. The stages were described as:

- Initiation
- Consolidation
- Integration
- Reflection

These stages are characterised by the type of questions that teachers ask themselves and others, and the type of teaching decisions they make.

At each stage, the teacher's ideas are modified and developed; the teaching situation encourages the teacher to ask new questions, and so ideas become formulated and integrated into new mental structures. Figure 9.1 illustrates the basic structure of each stage of the model. In the next four sections of this chapter, each stage will be described in greater detail, with supporting evidence drawn from the case studies. At each stage, the types of INSET which seem appropriate to a teacher at that stage are suggested.

In the present chapter, the stages are discussed in terms of a primary class teacher's development as an individual teacher of mathematics. It is also possible to discern stages that can be similarly

Figure 9.1 The basic structure of the model

characterised in other aspects of professional development. In Chapter 10, the model will be applied to organise the discussion of the role of the mathematics co-ordinator.

9.2 Initiation

At the first stage, *initiation*, the teacher is gaining experience of *what* to teach and *how* to teach it. The input to the teacher's thinking and decision-making is a jumble of ideas that were acquired in initial training: organisational ideas, mathematical content, educational theory, memories of teaching practice. Many of these ideas may not yet have been closely examined, and are only hazily recalled. The teacher's immediate needs are focused on the daily questions of '*What* shall I teach?' and '*How* shall I teach it?'. New teachers find this a difficult time, because it is necessary for them not only to think out for the first time the content of the year's work which they will teach, but also to experiment with organising it into learning experiences for the children.

Gradually, through experience, the teacher's knowledge becomes more integrated and better understood; a *personal teaching style* emerges. The teacher has found a way of coping with handling a class that is doing mathematics; a single limited but none the less personal approach to classroom situations has emerged.

At this early stage of professional development, teachers need help of a kind that they should be able to find within their own schools, on subjects such as the organisation of teaching groups, the use of equipment, and ways of presenting particular mathematical topics. Single-session INSET courses of the type that give 'instant' ideas on a topic such as arithmogons or other mathematical puzzles may be useful.

Although many teachers pass through the stage of initiation fairly quickly, this stage may be repeated in a modified form in later years of teaching when new content is taught, but by then the teacher has greater organisational skills to contribute to the new experience, and so is likely to return rapidly to a later stage of professional development.

The individual teacher and INSET

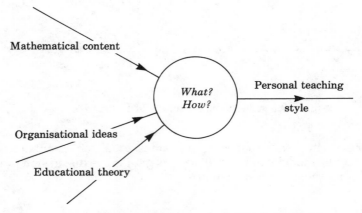

Figure 9.2 The stage of initiation

The researcher did not have many discussions with teachers who were at the stage of initiation — often they are not ready for INSET. As one teacher said as she looked back:

> ... perhaps for a few years you don't want new ideas ...
> (§8.5)

However, a young teacher at this stage came to the single-session course on a published scheme which the researcher monitored (§5.3). She responded to the course in a simple, straightforward way, feeling glad that she was already using some of the ideas put forward, and asking for similar sessions from other teachers, concentrating on classroom organisation and resources.

9.3 Consolidation

This second stage is mainly concerned with organisational teaching skills, and with a growing awareness of the pupils' needs. With increased experience, the teacher's attention turns to questions such as '*Why* does a particular teaching method work?' As the teacher's comprehension grows, a repertoire gradually builds up of different teaching approaches that are suitable for different topics and different teaching aims. The list of teaching strategies found in §243 of the Cockcroft Report (DES, 1982) may begin to be helpful in analysing the teacher's own classroom work. At the same time, the teacher experiences a growing awareness of children's individual ways of learning, and begins to relate teaching situations to the children's previous experience and their future learning.

The questions which teachers ask themselves at this stage are likely

The individual teacher and INSET

to be of the forms:

What shall I teach?
When shall I teach it?
How shall I teach it?
How do children learn it?

The introduction of *When?* questions indicates the broader outlook that has developed on the basis of experience and growing confidence. Personal teaching skills can now be applied in a variety of situations, unforeseen situations as well as those that the teacher had foreseen and planned for. The teaching has become more closely related to the needs of the learners, whom the teacher has begun to notice in more detail. This changing focus of attention, from the self to the pupils, is characteristic of this stage. At first, the teacher focuses on individual differences in the immediate classroom context, but gradually becomes more aware of age differences and progression through the school.

During this stage, the teacher comes to a much clearer *understanding of the curriculum*, at least in mathematics and other particular subject areas.

At this stage, teachers need help in setting their own work more clearly within the whole curriculum of the school. More extended experience of equipment and resources, discussion of various aspects of progression within the school's mathematics scheme, and the opportunity to work with children of more than one age group — these can all provide the necessary experience within the school to help the teacher at this stage. The teacher also needs to take part in a variety of short courses, on topics such as the use of a published mathematics scheme, or the use of structured apparatus in number work. Such

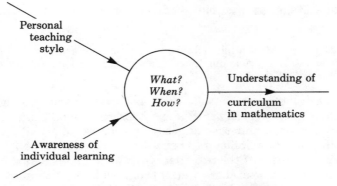

Figure 9.3 The stage of consolidation

courses, as well as helping the teacher to appreciate the progression of the mathematics curriculum, also give opportunities for mathematical discussions with other teachers, as do weekend residential courses on aspects of the mathematics curriculum, such as measures or the use of the environment. Teachers at this stage are likely to be able to make considerable use of ideas from these courses. The mathematics teaching journals are likely to become more useful, as the teacher is now more able to incorporate new ideas flexibly into classroom work.

Many teachers who seemed to be at the stage of *consolidation* discussed their experience of INSET with the researcher. One of those who had followed a one-term course had firmly established a variety of teaching styles in her classroom:

> If a child has a problem I now feel I KNOW how to help ... I am using more apparatus, and the children don't feel they are bad if they use it ... I realise how much children need to talk about maths ... and so I encourage children to come and talk about their discoveries. (§6.3.7)

Another teacher at this stage had needed to talk to the provider of a weekend course to reassure herself that the provider recommended a variety of teaching styles and not a single style based on environmental work (§4.2.4).

Other teachers at this stage stressed their awareness of the children's learning needs; one teacher went on a course because she wanted to find 'a realistic and practical progression for the child' (§6.3.3), while another had noticed that recently she could communicate better with the children (§4.3.3). Teachers discussed their reactions to INSET in ways such as:

> ... it is reassuring to hear that other people are doing the same sort of things, and you think 'Well, I am doing it right'. (§4.3.4)

Perhaps one teacher who found it interesting to talk to others about how they dealt with particular topics (§5.3.4) was further advanced within this stage, as was one who was ready for more substantial INSET:

> ... with the weekend, I felt stretched ... I feel that courses after school don't really stretch you enough, and I felt we achieved something ... we were working all the time ... (§4.2.3)

At an early stage of consolidation, however, teachers may find it very difficult to apply INSET ideas back in school. Some teachers rejected new ideas and teaching styles because of what seemed to them to be the inflexibility of their own environment (§4.2.8) or the unpractical idealism of the provider (§5.2.3 and §8.4). It is difficult to know whether these teachers would progress later, although we may suspect that some teachers, having consolidated a small range of teaching strategies with which they feel comfortable, progress little further, even when opportunities for development are placed directly in their path:

> ... when they did their final test their logical steps were so much better ... but I felt it was an isolated incident and once it was finished it was finished ... and we never carried it on ... (§7.3.2)

The experience of having to carry out problem solving each week for a term as part of the SPMPS project had not persuaded this teacher that it had anything to offer her.

9.4 Integration

A teacher's development in relation to teaching mathematics does not take place in isolation; gradually the teacher becomes much more aware of the many elements which cohere into a mathematics curriculum — schemes, equipment, progression, evaluation and assessment, the organisation of teaching, individual differences in learning. Additionally, the teacher is acquiring similar expertise in other curricular areas, and begins to consider the learning potential of interrelating different curricular areas. Discussion, analysis and reading make the teacher aware of a range of points of view, and challenge the basis for curricular decision-making.

The earlier questions do not disappear, but they are continually modified by '*Why?*':

Why is this topic included?

Why is this apparatus useful?

Why are groups an appropriate form of organisation?

Why does this topic precede that one?

It is a time of growing awareness of the decision-making element of teaching; the teacher becomes aware that decisions need to be consciously made, and that they need to have reasoned support to carry conviction, both with the teacher and with colleagues. Topics in mathematics are re-examined, and ideas may be broken down into parts and put together again in different ways, in order that the children's overall learning may be improved. This analysis leads the teacher to a more conscious awareness of generalised educational issues, both within mathematics and outside; in mathematics these issues may include the place of problem-solving, the nature and purpose of assessment and the role of calculators in the primary classroom. At this stage, the teacher becomes more confident and less subject to the whims of educational fashion, the anxiety to be 'doing it right' in mathematics teaching is dissipated, and the teacher

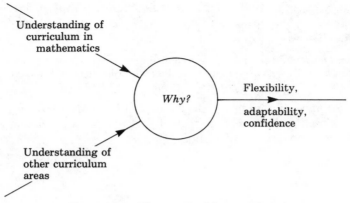

Figure 9.4 The stage of integration

becomes *flexible, adaptable* and *confident* as a teacher of mathematics.

Within the school, the teacher at the stage of *integration* is ready to take a full part in the discussion of issues concerned with the ongoing evaluation and development of the school's mathematics scheme, and to take up a leadership role. A similar involvement may also be found outside school — a discussion group, a group writing Guidelines for the LEA, a substantial course that gives time and support for reflection.

Many teachers who are at the stage of integration are to be found on substantial INSET courses. They come because they welcome the opportunity to engage in such activities as re-examining and thinking through a mathematical topic (§4.3.4); this may bring a growing realisation of complexity:

> ... it made you think of just how complex things are ... the more you work on maths ... the more you realise you don't know and the deeper you go the more holes you find. (§4.2.5)

One infant teacher visiting another school was particularly interested in seeing how mathematics developed in the junior years (§4.3.5), while study of the progression of a topic made another teacher much more aware of the interrelationships within primary mathematics: it 'was very difficult because the stages overlapped so much' (§4.3.5).

Comments made by teachers at this stage show an awareness of the wider curriculum and a wish to relate mathematics to it. One group on a weekend course related the mathematical topic under examination to the provision of rich classroom language, which would enable adequate language development, and tried to convince a colleague who was at an earlier stage:

The individual teacher and INSET

> ... we got talking about the real concept of addition ... and the language — how confusing it can be if there are so many different ways of saying 'add'. And ... you need to say practically all of them ... One person said 'I think I'll stick to one and then later I'll tell them the others', but the rest of us said 'No, you just keep saying it in all the different ways'. (§4.3.5)

Other teachers commented on a new awareness of the mathematical possibilities in the environment; several examples are found in §4.3.7, and this is perhaps the most vivid:

> ... someone brought a sunflower in ... and so we did a lot of estimating, and counting in tens ... and I probably wouldn't have used the full potential of that if I hadn't been on the course ... (§4.3.7)

For one teacher, now himself an INSET provider, an experience on a long course had provided a memorable insight, which had enabled him to integrate language and creativity with his mathematics:

> ... the first thing was an investigation ... I panicked — it seemed to be all words — no symbols ... I hadn't realised one could talk about maths ... I saw the answer overnight ... it was something I had been teaching ... it was a turning point for me ... in an investigation we can *create* (§4.4.3)

Such turning-points seem to be unusual; it is probably more common for teachers gradually to gain insights and so move on to a later stage in their thinking.

Teachers at the stage of integration are asking many *'Why?'* questions. One teacher went on a one-term course, hoping to find out 'why we teach what we do at infant school and to see the topic developed at junior level' (§6.3.3); another hoped for a deeper understanding of how children learn and of mathematical progression; she also wanted to find how to use and assess books and apparatus (§6.3.5). Concern about the basis for assessing textbooks and apparatus shows a higher level of thinking than is found in those teachers who are searching only for ways of using these tools in the classroom. Another teacher, when using the problem-solving project, fought her way through, with the children, to a deeper realisation of why mathematical symbolism provided an important — and economical — way of recording ideas (§7.3.3).

The importance of justification was stressed by one teacher as a major outcome from a long course:

> ... it gives you more confidence in your decisions, because you can justify it more ... because you know you can go to the right theory ... it makes you more critical of yourself as well ... (§8.5)

Thus we see that the fundamental confidence in one's mathematics teaching, which enables a teacher to undertake self-criticism and re-thinking and to live with the complexity of mathematics, is a characteristic feature of this stage.

The individual teacher and INSET

9.5 Reflection

Teachers at the stage of integration ask themselves and others '*Why?*' questions, but they often only seem to be seeking factual answers. The stage of *reflection* is characterised by a deliberate and meaningful questioning of what is being attempted in education, and thus by the asking of *Why?*' questions that seek value-based answers. Paradoxically, the integration achieved earlier has to be taken apart and reconstructed, so that each and every aspect of mathematical education can be reconsidered. The earlier questions are still of value — '*What?*', '*How?*', '*When?*', '*Why?*' — but the range of experience that the teacher now has available and the ability to employ it consciously in rethinking provide a full and rich context for each separate reconsideration.

Sometimes, the issues raised in rethinking cause an uncertainty that requires a partial return to an earlier stage. For example, a teacher may need to return to the stage of *consolidation* when reformulating an aspect of teaching style, or a return to *integration* may be required in reorganising the curriculum.

Another aspect of this stage, which might also appear to be a regression, is a reconsideration of the balance between content and process in mathematics. An awareness of process aspects of mathematics such as 'searching for a pattern' or 'making a representation in a diagram' will have come earlier. Now, however, the teacher is able to base decisions on a fuller consideration of the nature of mathematics and of its learning. What should be the balance between the teacher's teaching of 'other people's mathematics' and the pupil's creation of 'his own mathematics'? How can attitudes and skills be produced that will enable pupils to create their own mathematics? What should be the balance between setting pupils on a path that may eventually lead to their appreciation of mathematics as a 'pure' abstract structure, and stressing the 'applied' uses of mathematics in the environment and the sciences? The result of considerations such as these, both considerations such as the mathematical ones listed here and more general educational issues, is the development of a *personal educational philosophy*, which enables a teacher both to express beliefs in action in the classroom and to provide principled and well-founded leadership to colleagues.

Certainly, not all teachers reach this advanced level of professional thinking. It may be provoked in many ways: through considering the specific need of an individual child, through a challenge to one's view when sharing and discussing with others, or through a change of teaching environment, which offers an opportunity to compare different situations. These opportunities are, in fact, occurring all the time, but the teacher may not recognise them or accept them as a

The individual teacher and INSET

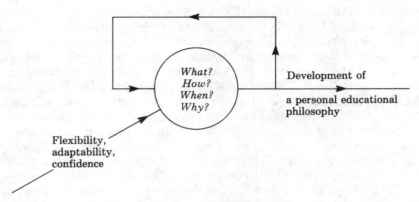

Figure 9.5 The stage of reflection

deeper challenge; when they are recognised and accepted, they lead to the development of an explicit and personal philosophy of teaching.

The loop in Figure 9:5 indicates the continual process of reflection and renewal that teachers at this stage experience. It might be thought that in order to gain an appreciation of education through mathematics it is necessary for the teacher to be very knowledgeable about mathematics, but the main prerequisite for accepting the challenge of this stage is enthusiasm for the subject and a desire to communicate this enthusiasm. Teachers with a good knowledge of mathematics may have this attitude, but this is not inevitably the case. Teachers without specialist knowledge of mathematics may catch the enthusiasm of others and then, while exploring the subject with their pupils, increase their own knowledge. At this stage the teacher is interested, first and foremost, in the educative value of mathematics.

Mathematical education is always moving on, as new circumstances such as the development of technology call for new emphases, and as new discoveries are made about children's learning and about teachers' teaching styles. Teachers at the stage of reflection are able to integrate new ideas such as these into their thinking and to convert them into classroom action. Thus, they often seem to be at the forefront of development in mathematics teaching.

When the stage of reflection is reached, a teacher needs *time* for thinking. Reading, discussion and study provide important inputs at this stage, so the necessary time may perhaps most easily be provided through a long course, probably for a diploma or a higher degree. Even if the study undertaken is not in mathematics, connections will be made and the teacher will reflect on the teaching of mathematics. There are other opportunities for reflective work in mathematics, both locally and nationally, in the working groups that are gathered together by LEAs, projects and the professional associations of mathematics teachers. At this stage, as well, it is natural for teachers to become INSET providers, both informally with colleagues and, more formally, as contributors to courses. For such teachers, it is important that they should develop the skills of the INSET provider: leading discussion, providing practical experiences for colleagues, teaching adults in an atmosphere of mutual respect and co-operation.

Perhaps not surprisingly, the few teachers who were reaching the stage of reflection whom the researcher met on and after courses, were largely found to be taking part in the challenging activities provided by real problem solving (§7.2) and by the SPMPS project (§7.3). Perhaps the very radical challenges provided by participation in these groups encouraged quite a number of teachers to reflect very deeply on what they were doing; others, however, rejected the challenge and either left the course or reverted to a previous teaching style as soon as possible. Those who had reflected on their teaching in a new way described their attitudes in such words as:

> ... on many occasions children were using maths much more creatively than I gave them credit for ... I became aware of their maths sense ... (§7.2.3)

> ... the democracy had to go on ... I feel it has built up a relaxed atmosphere for discussion ... and an acceptance of a changed teacher's role ... (§7.2.3)

One teacher had reflected on his previous style of questioning '... I load it so that I get the answer I want ... your face tells them ...'; he had worked his way through to an open acceptance of a changed teaching role:

> In the problem solving, your role does change from being the teacher with all the answers to a learning situation alongside them ... I'm still learning, because every time they are involved with the problems I see something else. (§7.3.4)

However, a few teachers who went on weekend courses were beginning to ask questions that are characteristic of this stage of thinking:

> When I go on a course ... I want to look again at what I am doing — justification, aims, objectives — we perhaps don't think enough about why we are doing what we are doing ... (§4.2.2)

> I think that kind of course is not to give you any answers, it is just to make you think and to give you ideas. Even if you don't use them immediately it stimulates your own thinking. (§4.3.7)

Probably, many teachers at the stage of reflection no longer attend courses of the types monitored by the project; they have become providers, as had a Head whom the researcher visited, attracted by

The individual teacher and INSET

the outstanding reputation of her school:

> ... when you open your mouth at a meeting or a course ... people catch the enthusiasm ... and I became caught up in talking at courses ... not just about the activities ... but perhaps more in a philosophical kind of way ... and we started a maths forum, and I did a term's course ...

> Then you find you're asked to run courses ... And I think there will always be ... among people that are interested in the teaching of mathematics ... these discussion groups ... and of course one is furthering one's own mathematics at the same time ... analysing how I work things out ... (§8.5)

Unfortunately, in few of the courses monitored did opportunity arise for the researcher to talk to teacher-providers about their ideas and philosophies, so this study does not give a great deal of explicit evidence about the stage of reflection, and teachers' values and needs at this stage.

9.6 The match between INSET courses and professional development

Quite a large proportion of primary teachers, from time to time, go on mathematics courses. Their hopes about what they will gain from courses are related to the stage of their professional development. When talking to the researcher, they often said that they needed stimulus — fresh ideas — the encouragement to think — a shot in the arm — the opportunity to talk to different people; they felt they were ready for a step forward.

> I wanted ideas and felt I lacked inspiration ... (§4.2.2)

> I went on the course purely because I thought I needed a bit of wakening up ... (§4.2.2)

However, mathematics is a subject about which many primary teachers feel anxious in the early stages of their professional thinking, and the theme of looking for reassurance that they were 'doing something right' in their mathematics teaching is one that crops up again and again in the comments of teachers at the stages of initiation and consolidation.

> When I started teaching, I didn't have anyone there to tell me if I was going along the right lines — if I was doing all right. You know, reassurance means a lot. (§4.2.2)

> ... it was encouraging when I found I was doing some things right ... (§6.3.5)

The next teacher was also probably searching for reassurance that she was 'doing it right':

> ... the keenness we all exhibited to look at children's work and compare it for standard and presentation with our own children. (§5.3.4)

Searching for reassurance may be the beginning of the growth of a more questioning attitude, and so may herald a move forward, or it may sadly lead to the closing of a mind, when a teacher becomes convinced of having found a way of 'doing it right' — or as right as others are — and so feels no need to look further.

Some teachers are further on in their thinking in other curriculum areas, and choose a mathematics course in the hope of overcoming this weakness:

> One should go in areas where one is weaker ... I haven't been to maths courses because I'm chary of maths myself. (§5.2.4)

> Maths was always a dreadful subject when I was at school, and this was one of the reasons that I wanted to go on that course ... (§4.3.3)

Sometimes, what happens on the course itself may have a negative effect, especially if the teacher has gone on what proved to be an unsuitable course for a particular stage of development. This may happen whenever the teacher's needs do not match what is offered:

> I am disappointed in the course — I thought it would have a lot of new ideas, it would be about the role of the maths co-ordinator ... but it's not. (§6.2.4)

There were other course members, at an earlier stage of thinking, who found this course interesting and satisfying. Mismatch between the course and a participant's thinking may discourage further participation or encourage complacency:

> I've heard people say — like I've said about this course — 'I know all that'. (§5.2.3)

There is a limit to the amount of fresh challenge people can cope with. On the two-day or three-day residential courses, total immersion in mathematics proved too demanding for a minority of the course members Other members, however, found these courses most satisfying:

> ... it was too much ... I felt ill on Monday ... It was not so much the course, but that one never stopped ... meal times ... at least I had a room of my own and did not go on discussing at night ... (§4.2.3)

For some course members, the conflict between the ideas presented on a course and the level of their own professional development may prevent new ideas from being put into practice. When that happens, excuses abound:

> Some of them were saying 'I think it is super and I would like to do this,

The individual teacher and INSET

but my Head wouldn't like it if I didn't do formal number work', and already they were envisaging problems ... (§4.3.3)

However, many teachers do gain from courses an appropriate fresh stimulus at their own level — a new idea to try with the class, a strengthening of existing convictions, or a challenge that pushes them to a higher level of thinking:

> I have done a lot of those ideas ... but I think you always doubt yourself after a while ...And when you try to get it over to other people, they have doubts ... I went there and I saw a teacher doing all these things, and it made me really feel 'Yes, it does work after all', and I came back revitalised. (§4.2.3)

> I have thought about how to help children understand — by going slower, by rethinking my questioning, by practical work ... (§6.3.5)

> I needed something new to latch on to ... and it was just a coincidence really ... there were all sorts of coincidences ... I was able to get on the course ... and she was an extremely good speaker ... and I needed something new to start thinking about ... (§4.2.4)

Different teachers find that different aspects of the same course stimulate them. On a weekend course, a teacher who was revitalised by a visiting teacher's presentation of her own work reacted strongly against the analysis of a topic, saying cautiously 'I don't think that was entirely useless.' Another teacher, however, found this analysis the most useful aspect of the course:

> ... thinking through a subject like length actually directed you to the way you are doing it ... If there was any follow-up to the course, I would really quite like to go through nearly all the topics and do the same thing that we did with length ... (§4.2.5)

The course ideas may stimulate the teacher's thinking, but eventually this should issue in classroom action. It is very difficult to judge how far a teacher is putting into practice ideas from the course; this was the most elusive aspect of the monitoring, especially as course participants could only be visited after an event, rather than their classroom procedures being studied beforehand. Occasionally, the teacher's Head was able to throw light on classroom actions, especially in the case of a long course:

> James says he has learnt a lot, but I don't see much change in his classroom. There are still compartments in his mind ... (§6.3.7)

Sometimes, a course member discovered, on reflection, that an idea that had been brought back from the course had died out in the intervening period:

> I was very interested in the games ... I did a bit of it after the course, and although it wasn't an ongoing thing — it tended to die off after a

The individual teacher and INSET

while — more from my fault than theirs ... I could see that there is a lot of potential in it. (§4.2.4)

In ideal circumstances, the INSET available would match the teacher's needs at each stage. When courses are provided in sufficient variety, this does seem to happen to a limited extent and in an unplanned way. Teachers try to choose courses that seem to them to be suitable to their needs as they perceive them, as we have seen above. However, a large proportion of the available provision of courses takes place at Teachers' Centres (see §2.1), and these courses are usually of the types that are most suitable for teachers at the stage of consolidation. Only very few teachers are able to attend substantial courses, which might challenge them to look more widely, so that their paths to the stage of integration were made more clear. This is not to say that the only path to the next stage is through an INSET course, but it certainly can often be an influence, by providing the stimulus that is needed. Courses that provide opportunities for the fundamental re-appraisal which challenges a teacher to the stage of reflection are even more rare, and a very small proportion of teachers have the opportunity to attend them. This is one of the greatest lacks in INSET at the present time. Many teachers who achieve a reasonable level of competence in their mathematics teaching are not enabled to have the challenge, combined with support, which would push their thinking to a higher level in the area of mathematics teaching, even if they achieve this in other curriculum areas.

9.7 Conclusion

The discussion in the previous sections may have given the impression that professional development is a linear process. This is not entirely the case. When a teacher is placed in a new situation — teaching a different age-range, working in a different type of school, being appointed to a post of responsibility such as that of a mathematics co-ordinator, becoming a head teacher, or starting to provide INSET — then even an experienced teacher at an advanced stage of professional development has to cycle back to the stage of initiation in the new role. However, experience and professional awareness makes this stage a short one, and the pace of development in the new role is usually faster. The inputs to the stages may now need to be described differently, but the questions of *What?*, *How?*, *When?* and *Why?* and the labels of *initiation, consolidation, integration* and *reflection* still remain helpful descriptions of the process through which a person goes in coming to terms with a new role. In

The individual teacher and INSET

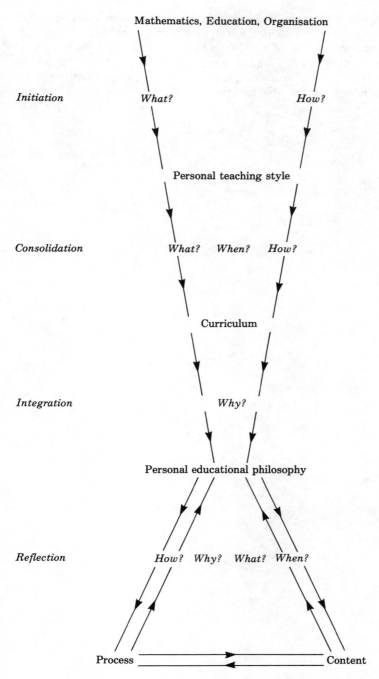

Figure 9.6 The stages of the model of professional development

The individual teacher and INSET

the next chapter, the model will be applied to the professional development of the mathematics co-ordinator in a primary school. Figure 9.6 gives a general overview of the stages of the model of professional development, emphasising that, when the stage of reflection is reached, the teacher continues to question, and to balance content and process in a holistic view of education.

CHAPTER 10

INSET in the Service of the School

10.1 Available INSET provision

Many of the available INSET courses in primary mathematics are focused on the needs of individual teachers, in the teaching of their own classes. This is certainly valuable in improving the quality of mathematics teaching in the teacher's class. It may also improve the quality of teaching throughout the school, if teachers are able not only to put ideas gained from courses into practice in the classroom but also to share course outcomes with their colleagues.

However, INSET of this type does not directly address the needs for improving the quality, progression and coherence of the mathematics curriculum and teaching of a primary school. External support can be provided for a school that is trying to do this, if it is possible to make an individual school the focus of INSET. Two methods are sometimes used: one is the provision of Advisory Teachers, who, in addition to working alongside individual teachers in their classrooms, often lead staffroom discussions on the mathematics curriculum, involving all the teachers in the school. The other method is the provision of INSET courses for individual schools, in which all the staff can take part, when the focus is on the individual school's needs. INSET of these types is expensive in human resources, and the second is still not at all common; these types of INSET were not monitored by the project, although the researcher did have some discussions with Advisory Teachers and with college lecturers who had taken part in school-based work (see Chapter 9).

In recent years, following the publication of the HMI Primary Survey (DES 1978) and the Cockcroft Report (DES 1982), there has been an increased emphasis on the establishment of a post of responsibility for mathematics in primary schools. Only by making a member of staff in each primary school responsible for most of that school's needs for INSET in mathematics can scarce resources be stretched far enough to reach and support all primary teachers. Consequently, there is a greatly increased need for INSET designed to help mathematics co-ordinators to do their work more effectively. Several courses specifically designed for this purpose were monitored and, on every other course that was monitored, some mathematics co-ordinators were among the course members. Additionally, the researcher had detailed discussions with several mathematics co-ordinators, and talked with Heads and Advisers about the role of the mathematics co-ordinator in the primary school.

10.2 The role of the mathematics co-ordinator

Promotion to the task of co-ordinating the mathematics teaching throughout a primary school makes considerable demands on a teacher, especially if that teacher's previous experience has consisted entirely of organising the work of a single self-contained class.

The appointment of a co-ordinator in a key curriculum area, such as mathematics, also necessitates changes in the thinking and the role of the Head teacher and of the other teachers in the school. In most primary schools, traditionally, the Head has provided most of the leadership in curriculum matters: perhaps using a democratic leadership style, perhaps not. It has been common, too, for teachers to be at liberty to think of their own classrooms as independent kingdoms, once the classroom door was shut; as one co-ordinator said when describing her problems: 'I have a teacher who may not adopt Nuffield' (§6.2.4). However, the complexity of the mathematics curriculum has increased considerably in the past twenty years, and it is no longer possible for each teacher to teach mathematics in a completely individual way, if the children are to have a satisfactory learning experience as they progress through the primary school. There is now a strong need for all members of staff to work as a team in their mathemtics teaching; this team needs day-to-day detailed leadership of a type which it is difficult for the Head to provide in all curricular areas. The necessary adjustments in formal structures within the school, in the structures of teachers' professional thinking, and in personal relationships between colleagues may not be easy.

The Cockcroft Report (DES 1982) listed a number of tasks for the mathematics co-ordinator:

In our view it should be part of the duties of the mathematics co-ordinator to:

- prepare a scheme of work for the school in consultation with the head teacher and staff and, where possible, with schools from which the children come and to which they go;

INSET in the service of the school

- provide guidance and support to other members of staff in implementing the scheme of work, both by means of meetings and by working alongside individual teachers;
- organise and be responsible for procuring, within the funds made available, the necessary teaching resources for mathematics, maintain an up-to-date inventory and ensure that members of staff are aware of how to use the resources which are available;
- monitor work in mathematics throughout the school, including methods of assessment and record keeping;
- assist with the diagnosis of children's learning difficulties and with their remediation;
- arrange school-based in-service training for members of staff as appropriate;
- maintain liaison with schools from which children come and to which they go, and also with LEA advisory staff.

(DES 1982, §355)

These are demanding tasks, and they necessitate that a newly appointed co-ordinator should advance to a new level of professional thinking, if the work is to be carried out effectively. However, a newly appointed co-ordinator will be at the stage of *initiation* in coming to terms with the role of curricular leadership. Much hard thinking, together with support both from within and outside the school, will be necessary before he or she reaches a level of professional development where all the tasks listed in the Cockcroft Report can be tackled in any depth.

We now discuss the topic of progression in professional development for a co-ordinator, in terms of the model of professional development detailed in Chapter 9.

10.3 The co-ordinator's professional development

10.3.1 Introduction

As they did earlier, when they were class teachers, co-ordinators progress through levels of professional development which we have described as *initiation, consolidation, integration* and *reflection*. Many teachers are appointed as co-ordinators when they are still at the stage of consolidation or integration in their own personal development as teachers of mathematics. There is, no doubt, much interplay between the stage of thinking a teacher has reached as a class teacher, and what that teacher is able to do in the role of a co-ordinator. However, it seems likely that all teachers return and recycle at least briefly through the previous stages in their new roles, as they come to terms with new tasks, responsibilities and pressures. In the case studies, there are several instances of teachers who are at a more advanced stage as teachers than they are as co-ordinators. The outstanding example of this is the co-ordinator interviewed in §8.6; he had been in the job for some time, and was in his third school as a co-ordinator. He believed passionately in practical methods in mathematics, but he still thought of the co-ordinator's role largely in terms of the organisation and ordering of apparatus, and worked behind the scenes to encourage his colleagues to make a fuller use of the equipment. The aspects of the role which are concerned with leadership were not yet part of his thinking; indeed, he explicitly saw himself as a foreman who provided the tools, while the Head's task was to work with teachers in the development of the curriculum.

10.3.2 Initiation

A co-ordinator at the stage of initiation is largely occupied in coming to terms with a new role. Co-ordinators need to start by acquainting themselves with the existing patterns of mathematics teaching in their schools, to expand their own knowledge of schemes and apparatus, and to get to the stage where they can answer other teachers' questions about schemes and apparatus, at least at a factual level.

The types of comments that teachers at this stage made to the researcher were:

> ... it's good meeting people with the same experience and the same school problems — it makes me feel they can be solved because others have done so. (§6.2.4)

> The Head asked if I would like to go because of my responsibility for infants, and because maths is not really my area ... (§4.2.2)

Some courses for co-ordinators, such as that described in Chapter 3, are explicitly planned to increase the knowledge base of the co-ordinators taking part; this course did this by discussing the apparatus needed in a primary school, by giving the teacher ideas for mental arithmetic, games and puzzles, and making sure that they knew of publications such as *Mathematics 5—11* (DES 1979) and other official pronouncements about the teaching of mathematics.

10.3.3 Consolidation

A co-ordinator at the stage of consolidation is now becoming more knowledgeable about the progression and co-ordination of mathematics throughout his or her own school. Fresh ideas are sought, and this leads to an interest in what others are doing:

INSET in the service of the school

> I didn't know what other schools were doing, or other people with posts of responsibility, and I just felt it would be a change to meet other people and see what they were doing. (§4.2.2)

The co-ordinator begins to share with other teachers ideas about mathematics and ideas for teaching mathematics, and encourages colleagues to do the same. In INSET for co-ordinators at this stage, plenty of scope for discussion is needed, so that ideas can be pooled and schemes of work from elsewhere examined. The courses described in Chapters 3 and 4 both gave plenty of provision for sharing and discussion.

10.3.4 Integration

Now the co-ordinator has reached a stage where the focus of development of the school's mathematics teaching begins to be grounded in his or her own beliefs, strengthened and tempered in discussion with colleagues. Realisation has grown of the strengths and weaknesses of colleagues' mathematics teaching, and thus of the need for school-based INSET. Awareness of children's learning difficulties makes it necessary for the co-ordinator to lead the school to review its way of teaching mathematics, and the co-ordinator is ready to lead colleagues in revising the school's scheme, or to focus INSET on some area of difficulty in teaching, such as place value. The co-ordinator now sees more clearly how to integrate mathematics into the school's general curriculum principles, and to integrate it with the teaching of other curriculum areas.

A co-ordinator who was beginning to reach this stage said to the researcher:

> ... I have done a lot of those ideas ... but I think you always doubt yourself after a while, as to whether it is really worth while. And when you try to get it over to other people, they have doubts — and they say 'Are you sure these ideas actually work?' ... I went there and I saw a teacher doing all these things, and it made me feel 'Yes, it does work after all', and I came back revitalised. (§4.2.3)

At this stage, the co-ordinator's own INSET needs are wide-ranging: the inspiration that is needed to strengthen resolve, a detailed study of the progression of topics, personal experience of problem solving, investigations and environmental work, assistance in structuring a scheme of work. A long course may provide both the impetus and the time for reflection needed.

10.3.5 Reflection

At this advanced stage of professional thinking, the co-ordinator's attention turns even more strongly to ways of helping and influencing

INSET in the service of the school

colleagues, and the need for great skill and sensitivity in personal communication is realised. The co-ordinator is now ready to consider and implement strategies for change such as working alongside colleagues in their classrooms.

> I've gone through this problem of how you actually approach people to get work done ... How do you talk to your Head ... how do you set about in-service training within your own staffroom ... it's a very knotty problem...
>
> ... the thing about the course was 'How am I going to pass investigations on to other people?' (§4.4.5)

By now, further study will probably have brought understanding of general principles in the learning and teaching of mathematics, and the co-ordinator will be able to apply these general principles in specific situations in the school; he or she will also be able to diagnose the strengths and weaknesses of the school's approach to mathematics teaching, and to plan the directions in which it should be encouraged to move in the future.

There is probably no longer much need for formal INSET, although the co-ordinator will be well aware of the need to keep up to date in thinking about advances in mathematical education. Much of his or her own INSET may now come through membership of working groups that are set up locally or nationally to carry out some developmental task; the teacher also will almost certainly have become an INSET provider in a larger forum than the individual school, and so will continually hone ideas in discussion with colleagues from different areas of the educational scene — secondary teachers, Heads, Advisers, lecturers.

By now, probably, the co-ordinator is no longer a co-ordinator — promotion will have come again, and the experience gained in the co-ordination of mathematics teaching will be pressed into service in the teacher's new role of Head or Deputy Head, Adviser or lecturer.

10.4 Feedback after courses

The interplay between INSET courses outside the school and work within the school is a delicate one, not only for the co-ordinator but also for all members of staff who go on courses.

In the last analysis, the purpose of INSET is to raise the quality of teachers' professional thinking, and so to have an effect on classroom practice. However, it is all too easy for a teacher to attend an INSET course and then to return to school as if nothing had happened, and to carry on teaching exactly as before. When this occurs, it may well be that time is necessary before the teacher's new thinking can issue in new action, and the action will come later. Alternatively it may be that the course did not produce any new thinking for that teacher; this is not necessarily a criticism of the course — the teacher might not have been at the right stage of professional development to profit from a particular course. Another reason why a course might not issue in action on the teacher's return to school is that, when the teacher again became immersed in the school situation, the pressures of everyday life caused the new thinking that was beginning to occur to be rejected, because it seemed unrealistic or too difficult to put into practice in the teacher's particular environment.

If new thinking has tentatively begun, the attitudes and support that the teacher meets on re-entry to school can be crucial. Some schools try to ensure that a teacher articulates what has been gained from a course, and that others also profit from that teacher's experience, by asking course-goers to report back to a staff meeting. The researcher did not attend any of these reporting-back sessions, but several teachers commented to her about the difficulty of reporting back to colleagues.

> ... so much was said on the course ... it would be difficult to transmit it. I think I gave them the feeling ... but it's difficult, isn't it ... (§4.2.6)

Even if the actual reporting back is not very successful, the value for the teacher of putting what happened into words for colleagues is not to be underestimated. It is also possible to think of strategies that might make the value of reporting back greater. Course providers might give some thought to assisting the teacher who is trying to report back, by providing a handout to summarise points that could appropriately be reported on. Other course handouts may also be valuable, and spare copies can sometimes be put to good use within a school. The course tutors of the one-term course described in §6.3 tried to support feedback by arranging an exhibition of course-members' work, to which colleagues from their schools were invited. This was intended to provide a focal point for discussions, which could be followed up later in school, and in some cases it was effective in doing this.

It may also be possible for the teacher's own school to assist in feedback. If the co-ordinator is able to discuss a course with the teacher before the staff meeting at which feedback takes place, consideration can be given to points (if any) from the course that can be followed up in school, with the co-ordinator's support; thus, in selected cases, a reporting session can be used as the first phase of a new initiative in the school. For example, a teacher's attendance at a course on calculators may lead to a consideration in a staff meeting of the place of calculators in primary mathematics, and then to experimentation within the school.

Sometimes, several teachers from a school are able to attend the same course. This may have considerable advantages, for the teachers may be able to support one another in carrying out actions that result from the course. Several teachers also form a more powerful force than a single individual, as they present the course ideas to their colleagues. However, the two teachers from a school who attended the course described in §4.2 found that, although they supported each other, they had difficulty in communicating their enthusiasm to their colleagues.

> ... We found that they were too tired and we were full of enthusiasm, and we thought ... we'll wait and see what the response is, but I think they were just too tired to receive it. (§4.2.6)

At least, however, these teachers were able to try to work out their ideas together, and gauge what the children's response to them was; this may be the best way in a school as unsupportive as this one seemed to be:

> ... when you try to apply it in your own situation you have got to find your own solutions ... It was lovely to see all those pieces of work laid out there, but there would be nowhere where we could actually put anything like that ... you come up against people who don't particularly want things there ... (§4.2.8)

Ideas for development that emanate from two people have more chance of growing and taking root than those put forward by a lone voice; it is not surprising that the ideas brought back by individuals are often lost in the difficulties of persuading colleagues and in the practicalities of the school situation.

10.5 Re-entry

A problem which is related to that of feedback is re-entry to school after an INSET course. Feedback is concerned with disseminating the ideas from a course more widely, while re-entry is concerned with the teacher's own experiences after the course. Re-entry is especially difficult after a long course, but re-entry problems may occur even after a single-session course. For the course-goer, the experience of the course is usually an individual one, not shared with other members of staff; if it changes the course-goer in some way, then a process has to take place in which some balance is achieved between assimilating the course-goer back into the school situation, and adapting the school situation to the course-goer's new ideas.

A long course, when a teacher is away from school either for a term or more, or intermittently over an extended period, may have a profound effect on a teacher — 'It was a turning-point for me'. The supportive atmosphere and shared activities of many courses are also very different from the daily routine of teaching. In all, return to school after a long course is a difficult time for a teacher; the problems of re-orienting in such a way that the benefit of the course is not lost may be great, and sensitive support is needed from Head and colleagues.

It is probably better for a teacher to adopt new content and teaching styles gradually, rather than to rush into them; after a one-term course, a teacher who moved from formal class teaching to an organisation based on four groups found that her organisational skills did not yet enable her to cope with more than two groups (§6.3.6). This engendered some sense of failure, and she seemed to have little back-up from others in the school as she attempted to make these changes.

The most important need on re-entry is for support from the school, in an atmosphere in which sharing is accepted as normal, so that the teacher's new ideas can be implemented with the understanding of colleagues. The teacher also needs to think about the problem of disseminating new knowledge — 'Martin's approach is not threatening to other teachers' (6.3.7). After half a term, one teacher from this one-term course felt ready to lead a staff meeting, but another learnt she had to 'be careful as they are touchy'. Indeed, at the end-of-course display of work before they returned to school, some teachers only gave brief and apologetic explanations of their work, and had an obvious desire not to appear too knowledgeable and enthusiastic in the eyes of their colleagues. These teachers were already aware of their need to be re-assimilated, at least on the surface, into the style of working of their own schools with their familiar colleagues. It would not be surprising if their gains during the course were rapidly dissipated on re-entry to school.

The key figure in setting the atmosphere of a school is the Head, and we now turn to ways in which Heads can influence the use that is made of INSET in mathematics in their schools.

10.6 The influence of the Head

The most propitious situation for developing mathematics teaching in a primary school is one in which the whole staff are able to work as a team, sharing their ideas and their planning, and always being ready to consider fresh approaches, to experiment with them and to evaluate them. In such an atmosphere, confident and knowledgeable teachers can move forward, and teachers who are lacking in confidence in their mathematics teaching are able to express their difficulties without worry of losing face or being derided; they can then benefit from the help of more confident and imaginative colleagues. Only the leadership of the Head can produce a situation in which all the teachers learn together and work together in this way.

Some Heads are themselves not confident about an approach to mathematics teaching; this need not be a handicap to development, provided that it is clear to everyone that development has the Head's whole-hearted involvement, backing and support, and that the Head is thoroughly concerned to work alongside colleagues to improve the mathematics teaching in the school. One example in which this did not happen is found in §8.3; the Head called in an outside agency to work with her teachers, but she did this without adequately thinking out the preparation needed, or being prepared to give sufficient practical backing. Two of the three teachers involved were not at a suitable stage to be able to build on the visitor's ideas, and practical follow-up and support had not been thought through; the visitor summed up his impressions to the researcher:

> ... there was very little apparatus for weighing and so on ... the Head is very inclined to art ... you know Heads spend their money on what their interest is ... so I shouldn't think it led to them getting more maths apparatus ... (§8.3)

Other Heads, however, are able to provide the right support at the right time to enable teachers to work out new approaches and developments in their mathematics teaching. One teacher was encouraged to feel how much her new teaching style was valued, as her Head asked her to repeat the movement work she had done on pairs for the older children to watch (§4.2.4). Another Head was able to lead a teacher to see that work on attributes that she had previously done in a formal way with structured apparatus could be done equally, or more, effectively in an informal style, using the children's environment as a starting point (§4.2.4).

Probably, the teacher who needs the most support in mathematics from the Head is the mathematics co-ordinator. The Cockcroft Report states that:

> Good support from the head teacher is essential if the mathematics co-ordinator is to be able to work effectively, and some modification of the co-ordinator's teaching timetable is likely to be necessary in order to make it possible to work alongside other teachers.
>
> (DES 1982, §357)

This modification of the co-ordinator's timetable is a measure of the Head's visible commitment, as it often involves the Head in personally taking over the co-ordinator's class. However, much more than this is needed if the co-ordinator is to develop so as to be able to work effectively; working alongside other teachers is not the first job that a teacher who has recently become a co-ordinator is likely to be able to do, and the Head's commitment and willingness to support the co-ordinator, and to delegate important tasks, needs to be demonstrated from the start. The Head may have to help the co-ordinator to see how the job can develop; the co-ordinator who discussed his task with the researcher (§8.5) had a very limited view of his role, and relied too much on the Head to ensure the development of his colleagues' mathematical thinking:

> We have got a new scheme to watch over — now I would see that as the Head's job, not the co-ordinator's . . . if teachers are in difficulty, I would see them turning to the Head . . . (§8.6)

On the other hand, it may be that the co-ordinator needs to nudge the Head gently into seeing the possibilities, and into encouraging the staff to move forward:

> I talked to the Head . . . I showed her all the leaflets we had on the course and said 'Perhaps you'd like to read them' . . . she's not really a maths person — she's a language person . . . (§4.4.5)

The Head may also have to work to foster changes in professional relationships among the staff; these may be necessary if the mathematics co-ordinator is to provide leadership to colleagues in one curricular area, while working under the leadership of others in the remaining curricular areas. All the teachers need to come to respect and enjoy each other's special strengths. The Head's leadership in doing this can be crucial; on one occasion, one of the writers was invited to attend a staff meeting in order, as the Head explained afterwards, to strengthen the whole staff's conviction that the co-ordinator's views on the importance of practical mathematics not only had the support of the Head, but also had the enthusiastic backing of 'expert opinion' in primary mathematics.

Edith Biggs, in her account of her action research project on in-service work in primary mathematics (Biggs 1983) makes absolutely clear the Head's central role in maintaining commitment and momentum. The project school that was the most effective in changing its mathematics teaching did so in the face of serious difficulties:

> [It] had a cumulative staff turnover of 60 per cent. There were two changes of co-ordinator in 1980, but the head's interest and determination enabled the changes to be sustained while she herself trained the third young co-ordinator. Moreover, during this period, a comprehensive mathematics checklist was prepared by the head and all the teachers to use with individual children.
>
> (Biggs 1983, p. 184)

CHAPTER 11
Providers and INSET

11.1 Identification of needs

All the courses monitored by the project were initiated by agencies other than schools; these included Advisers, Wardens and planning groups of Teachers' Centres, and colleges and the Open University. Only in the case of the series of visits to schools (§5.3) did the initiative come from some of those teachers who wanted to participate. In other cases, a need was identified by someone other than a teacher, often an Adviser, and a course was run to fill that need. Sometimes Advisers ran courses themselves; in other cases, college lecturers or teachers were the course providers.

This method of organising course provision raises the question of how close the match is between teachers' own perceived needs and interests, and their needs as they are perceived by those who, although in close contact with teachers, are not themselves classroom teachers. In most of the courses monitored, there was a fairly good match between what was provided and the teachers' needs as they came to realise them during the course, even though many of the teachers could not have identified in advance their need for topics such as environmental mathematics or real problem solving. It is an important facet of INSET to make teachers aware of needs that range further than those they might themselves identify; however, a balance certainly needs to be kept with teachers' own perceived needs.

11.2 Planning and advertisement

In Chapters 9 and 10, it was suggested that teachers at different stages of professional thinking have different INSET needs, and that they tend to turn to different sources of provision when they decide that they would like to go on a mathematics course. Teachers in the early stages turn more to Teachers' Centres, and generally attend short courses; teachers at a later stage of thinking often look for more substantial courses, such as weekend residential courses and long courses in institutions of higher education.

This makes it very important that the course advertisement should identify as clearly as possible the type of course that is to be mounted, especially if a course of that type would not usually be expected from a particular provider. Will the course provide instant ideas for classroom use in the immediate future? Or will the speaker expound a philosophy of mathematics teaching? Both types of course are found at Teachers' Centres, for example, and they satisfy the needs of different teachers. It is not always easy to discern from the advertisement which type of course is intended — indeed, sometimes even the course organiser may not know which it will turn out to be. The course described in §5.4 turned out to be philosophical in nature, although the title and advertisement stated that it would be concerned with a particular published scheme. The teachers who appreciated the course most were those who saw behind the advertisement and came with a realistic expectation:

> I had heard most complimentary reports of the speaker ... I expected little more than to hear an experienced educationalist ... and gain advice and benefit by listening to him ... I found the lecture refreshing, entertaining and fulfilling. (§5.4.2)

Those who took the advertisement at face value were disappointed:

> I was disappointed that the talk was not more specific about certain items.
>
> I thought it was rather general in its content ... (§5.4.2)

The course organiser is certainly able to shape the content and format of the advertisement, but it may be difficult to ensure that the actual course content matches the advertisement. In the intermittent course for co-ordinators described in §6.2, the advertisement was well matched to the perceived needs of co-ordinators, but, although all the topics advertised were dealt with, the balance between them reflected the convictions of the providers and the areas of their strength, rather than the expectations of the course members. Only a proportion of the course members found during the course that the balance provided suited their needs, and there was consequently some disaffection. Perhaps the organisers might have made more use of outside speakers to deal with areas outside their own experience.

However, when outside speakers are brought in to supplement what the providers themselves can give, the organisers may have little control over how an external speaker interprets the brief given. In fact, the experience of the present writers suggests that more might sometimes be done to brief one-off speakers. A copy of the course advertisement is always useful and, for a longer course, a discussion (perhaps by telephone) on how a particular session is intended to fit into the total course structure may enable the speaker to slant the session in the hoped-for direction.

An issue that is related to that of advertising the course is that of selecting among the applicants, if this needs to be done. If the course provider does any necessary selection, it can be done in accordance with the provider's clear vision of the type of teacher to whom the course will be suited. Usually, this is more possible in the case of long courses, when applicants can be interviewed. However, in the case of the intermittent course for co-ordinators (§6.2), the selection was done by the Mathematics Advisers of the LEAs which commissioned the course. This division of responsibility is only likely to be successful if the selectors and the providers have a shared idea of the focus and aims of the course. Otherwise, mismatch between the course members and the course is only too likely. For over-subscribed short courses, arbitrary methods, such as the first to apply or taking only one teacher from a school, have to be used; a careful and explicit course description in the advertisement can make a degree of satisfactory self-selection more likely.

11.3 Course timing and structure

Many course organisers have little control over the timing and structure of their courses; a residential centre may only be available for a particular number of days at one particular time of year, or the provider's other commitments may force a one-term course to be held in one term rather than another. It is usually impossible, for a variety of reasons, to run a course on the ideal day, at the ideal time of day.

However, the timing and structure of a course certainly interact with its content. The experience of some of the teachers on the course described in §4.2 suggests that if it is an ingredient of the course plan that participants should report back to colleagues in their schools, July is not the best time to hold the course. However, July may be an excellent time for a course whose main thrust involves teachers in thinking about their planning for the new school year. Thus, priorities in course aims that are seen by the provider may well conflict with the timing and structure available to implement these aims.

Courses usually seem to their providers to be too short, and the temptation to make a plan that crams in more material than the available time will hold is very strong, as the provider of the three-session course described in §5.2 found to his cost. The collapse of his original plan led to an excessive slowing of the pace, and frustration all round. Because course time is so scarce and valuable, teachers usually appreciate it greatly when all the available time is fruitfully filled, even if the result is very intensive; some of the teachers on a weekend residential course made it clear that, having given up a precious weekend to the course, they appreciated the filling of every available moment: 'We achieved something ... we were working all the time' (§4,2.3).

11.4 Discussion and practical activities

Among the valuable features that courses may include, good opportunities for discussion are regarded very highly by many course-goers; on the other hand, if the programme incorporates too many discussion periods, it may suggest that the work of the course was not thoroughly well-prepared (§6.2.6). To get the most out of it, discussion needs sensitive and tactful leadership and usually flows well when a thought-provoking stimulus is provided. However, the skills of planning and leading discussion are not necessarily among those which new INSET providers have acquired from their previous experience, and many providers need to acquire these skills on the job. As an example of tactics that can be useful to a provider who is not a skilled chairman, *group activities* can provide an opportunity for stimulating and well-focused discussion. On a weekend course, the activity of sequencing a topic (§4.3) ensured that informal discussion of curriculum issues would take place; the providers engineered this discussion by their choice of the sequencing problem which the course members tackled.

Even when the stimulus provided for discussion is extremely provocative, as in the case of the classroom visits described in §5.3, the discussion may fail to take off, for one reason or another. In the situation described, course members were afraid of appearing to criticise their hosts; if a sensitive leader had been available to chair the discussion, it might have been possible to focus questions and discussion towards less threatening areas, such as the teacher's reasons for her choice of classroom organisation, and the advantages and problems of that type of organisation. A focused discussion such as this would not have threatened the host teacher, who would then have taken the role of the expert explaining her style of organisation.

In the planning of an INSET course, some messages may best be conveyed covertly, rather than by direct statement. For example, long courses often include the learning of mathematical topics that are likely to be unfamiliar to the course members; experience of the style of learning used on an INSET course can often help teachers to develop their own styles of classroom work, because they realise that if they themselves benefit greatly from practical group activity and its accompanying talk when they are learning mathematics, then children's need for this style of working must be so much the greater. As one teacher said: 'I realise how much children need to talk about maths, just as I did on the course' (§6.3.7).

There are other ways in which one area of content can be used to carry covert messages that may not be acceptable if they are conveyed directly. For instance, many teachers whose own mathematical understanding is weak would not be confident enough to go on a course that overtly claimed to teach them mathematical ideas; they may be more willing to attend courses that deal in a practical way with the teaching of mathematics in the primary classroom. Through the study of apparatus and materials, and through using them in their own classrooms, teachers can often come to a deeper understanding of areas of mathematics in which their knowledge was superficial.

When unfamiliar apparatus or games are introduced to teachers, hands-on experience is essential, and adequate time must be allowed for this; it is most counter-productive if the reaction to the experience is like that which a teacher described to the researcher:

> People have said to me that they wouldn't dream of ever using Dienes' apparatus because of an experience they've found to be very unhappy ... [on a course] for a couple of hours.(§8.4)

Time for experiment and discussion in an unhurried and unthreatening atmosphere is necessary if new ideas are to be digested and incorporated into the teacher's thinking.

11.5 Handouts

In conversation with the researcher about several of the courses monitored, teachers stressed the importance of receiving good handouts from a course. The handouts given out by the provider often form the only concrete residue from the course, and serve as a reminder for course members when the course is over; they supplement the teacher's own notes (if they have taken any) and any written work produced for assessment. Perhaps the course members will never again consult the handouts after the course has finished, but, if there are no handouts available for later consultation, or if the handouts do not communicate easily when their original context has faded from memory, then a valuable source of support for the ideas promoted in the course is lost.

The project identified several ways in which providers can plan for their course handouts to be used after the course has finished. These include:

- guidance to course members for follow-up work;

Providers and INSET

- information about sources of resource materials;
- a structure for the dissemination of course ideas to the course-goer's colleagues;
- material that colleagues themselves can read.

The handouts provided on the weekend course for co-ordinators (§4.4) were of a notable standard of excellence and provided examples of all four types of use. The checklist of aspects that should be included in a scheme of work was a good example of a handout that gave teachers guidance and structure for future work. A list of starting points for sources of investigation provided for future sources of resources, and all the handouts were helpful in dissemination, and suitable for others (particularly the co-ordinator's head) to read.

The project also found that handouts were being used for a variety of purposes during courses:

- to provide stimulus for practical activities;
- to summarise the previous course session;
- to provide key extracts from official publications as a focus for discussion.

When courses are run by staff of higher education institutions, back-up is always available for the production of handouts. In the case of short courses mounted elsewhere, it is not always clear how far visiting providers can expect the necessary secretarial services. Certainly, this type of support is necessary, if providers are to ensure that their work is supported by high-quality handouts.

11.6 Qualities of providers

Providers may structure a course carefully, provide interesting and stimulating activities, and distribute excellent handouts, but often what remains in a teacher's mind is a general impression based on qualities of personality, which the provider may feel are not very much under his or her own control: 'She was an amazing woman ... she was really full of enthusiasm ...' (§4.2.3). The enthusiasm of the provider's presentation is certainly greatly valued by course-goers. When teachers come to a course session after a tiring day in school, the urge to sleep through an unenthusiastic presentation can be overwhelming; on the other hand, an enthusiastic presentation can give teachers the necessary impetus to attempt something new after the course. However, the quality described as enthusiasm is probably a compound of the provider's conviction of the value of the subject matter, so that its communication to teachers is of great importance to the provider, in addition to careful planning, good illustrative materials and a well-structured presentation.

The project was interested to note that some courses that were described by members as excellent were run by providers who would not have thought of themselves as powerful speakers. Their success often came from the quality of the activities that they provided: 'I had a whale of a time ... we were doing different projects ...' (§4.3.5). Those projects had been planned by the Advisory Teacher, who for a long period had worked with the teachers who actually led the project groups, in order to build up the teacher-leaders' confidence and style of working, and to help them to develop to a stage where they could themselves become successful INSET providers. The Advisory Teacher's overt contribution to the session was merely to provide an introduction and summary, but without her high-quality work behind the scenes, the session would have been impossible.

Contributions from serving teachers are often greatly valued by course members; these providers speak on the basis of ongoing experience of classroom work, and they can often show examples of children's work on the topics they discuss. They carry conviction that what they recommend does actually work in practice. However, successful experience in the classroom does not always translate immediately into equal success as an INSET provider. As one course-goer put it:

> ... if the lecturer is a teacher, maybe they forget they are with adults rather than with children, and they put the material across in the same way as they would with children. (§5.2.3)

The model of teacher development, discussed earlier, applies not only to teachers but also to INSET providers, whether they are teachers, Advisers or lecturers. They, too, go through the stages of initiation, consolidation, integration and reflection in the role of an INSET provider, as they gain experience and come to think more deeply about the task of providing fruitful INSET for primary teachers.

There is little training, or even pedagogic literature, available to help a new INSET provider. It is hoped that this book will make some contribution towards building up a body of literature focused on the needs of INSET providers. However, probably one of the most helpful experiences by which a provider can develop is to be fortunate enough to be able to work in a team, taking part in the planning, provision and evaluation of courses alongside more experienced colleagues.

CHAPTER 12

Conclusion: INSET as an agency of change

12.1 Introduction

In the last three chapters, we have summarised some implications of the project's monitoring of INSET courses in primary mathematics for three parties concerned in the joint enterprise of educational development:

- the individual teacher
- the school as a whole
- the INSET provider

The project has identified a number of ways in which it may be possible to make INSET in primary mathematics more effective. It has also, more significantly, put forward a model of teacher development, suggesting that, on their route to full professional maturity, teachers pass through identifiable stages, which have been labelled initiation, consolidation, integration and reflection. This model has been derived from the project's experience, but it certainly needs further study and validation.

12.2 INSET as an agency of change

For an individual teacher, INSET can be a potent force for the development of that teacher's professional thinking in primary mathematics. It can also be a potent force in encouraging a whole school to achieve a united aim in its mathematics teaching, and to base the process and content of that teaching on surer foundations.

Conclusion: INSET as an agency of change

Indeed, INSET is *the* agency of change in the educational system. If there were no INSET, there would be no opportunity for teachers to discuss with colleagues the teaching of mathematics (or of any other curricular area — but we mention mathematics, because mathematics teaching is the subject of this report) or for teachers to take into their thinking any ideas for development other than those that grow in their own minds; nor would there be any opportunities for teachers to learn at first hand about new apparatus, books and other resources.

Among the varied types of INSET, school-based INSET is one of the most important. This takes place when the whole staff of a school, or a group of them, come together to discuss and try to solve the problems of their mathematics teaching; it enables all members of a school staff to take their own part in shared thinking and united development of their mathematics curriculum and teaching. School-based INSET is also cheap or, rather, it is cheap if only a low (or zero) level of external support is provided to each school. If a good level of external support were to be provided to every school, school-based INSET would be very expensive indeed. Even in an ideally well-resourced educational system, school-based INSET would have a very important part to play — it is the means by which a school achieves a common set of goals for its mathematics teaching and implements them in a coherent and suitable mathematics curriculum.

However, on the basis of the study that the project has undertaken, we are convinced that school-based INSET should never be the *only* form of INSET on which any school should rely. There are several reasons for this: school-based INSET relies almost entirely on the ideas already available within the school's own thinking, and, unless external influence is brought to bear, either by a visitor or by a teacher who has taken part in other INSET, little real development may occur. Moreover, many primary teachers lack confidence in mathematics teaching, and so find it extremely difficult to develop their style of teaching; new ideas and advances in mathematics teaching do not easily arise spontaneously and grow within every school. External influence and support are most urgently necessary, and these can only be provided realistically through INSET courses, not by external suppport to each individual school.

Among the courses required to support school-based INSET are courses that help the participants to *focus* on the varied needs of their schools. In the primary schools that the researcher visited, the project found no evidence of any systematic planning for meeting the INSET needs of the school. In 1978, the Induction and In-Service Training Subcommittee of ACSTT (DES and Welsh Office 1978) identified four practical steps that any school could take to plan its own INSET programme:

Conclusion: INSET as an agency of change

Identify the main needs.

Decide on and implement the general programme.
Evaluate the effectiveness of this general programme.
Follow-up the ideas gained.

The project found no evidence of this type of planning. INSET still seems largely to be thought of as an individual task, to be undertaken only when an individual feels the need. It is part of the task of the mathematics co-ordinator to arrange school-based INSET, but it might also be thought to be part of the task of the Head and the co-ordinator together to plan to satisfy the INSET needs in mathematics of the school, and to ensure that the programme is implemented.

12.3 The level of provision

The project was not able to examine whether the present level of INSET provision is sufficient to make it possible for at least one teacher from every primary school to take part in the type of substantial INSET course that would enable him or her to develop professionally in order to take a leadership role in mathematics teaching. However, it seems extremely unlikely that this is the case. A study of the level of take-up of INSET in primary mathematics would be fruitful in gauging the appropriate level of INSET provision that would be necessary if the teaching of mathematics in *all* schools is to be influenced.

We have seen, too, that the provision of substantial courses needs to be supported by a good level of provision of short courses at Teachers' Centres and similar venues, as well as school-based INSET; in the early stages of their professional development as teachers of mathematics, teachers are not able to draw maximum benefit from a substantial course and short courses are necessary at this stage. If such courses are not available on a regular basis, then teachers are unlikely to start to progress in their thinking about the teaching of mathematics beyond the stage of consolidation, the stage at which they established some sort of style of mathematics teaching, however effective or ineffective that might be.

In its chapter on 'In-service support for teachers of mathematics', the Cockcroft Report noted that:

> The evidence which is available to us suggests that at the present time sufficient money is not being spent on the provision of in-service training and that in some areas the position is worsening. Unless the Secretaries of State take effective action in this field we do not believe that sufficient resources to improve the quality of mathematics teaching will be made available.
>
> (DES 1982, §765)

Conclusion: INSET as an agency of change

The Secretaries of State have taken some action in this field, and benefits are flowing from it. Because of the intervention of the Secretaries of State (DES 1983), there is now some provision for supply cover to enable mathematics co-ordinators in primary schools to undertake substantial INSET courses in school time. This is a considerable advance, although the scale is not at all large. Indeed, the actual scale of provision under Circular 3/83 was not even as large as the modest scale intended; some LEAs were unable to take up their allocation, because no corresponding provision was made to enable colleges to mount the additional substantial courses needed. However, a beginning has been made, and it points the way to further development.

12.4 After Cockcroft, what next?

In an age in which the pace of technological change quickens daily, and dependence on mathematical understanding increases all the time, it is necessary that the children who attend every primary school in the country should be enabled to receive a high-quality mathematical education. It is difficult to estimate the scale of the changes that are needed to achieve this end, but some indication can be drawn from a survey of the use of published mathematics schemes in classes containing 11-year-old pupils; this survey was undertaken by the APU in 1982 (Foxman 1984). The survey found that the most widely-used scheme was Alpha/Beta, which was used by nearly one-third of 11-year-olds. This scheme was first published in the late 1960s and, although it has been superficially updated since, its approach to the teaching of mathematics is still one that was in vogue a quarter of a century ago. It does not enshrine any of the substantial developments in primary mathematics teaching pioneered at about the same time by such workers as Edith Biggs, HMI, and by the Nuffield Mathematics Project. Further responses to the APU survey indicate that, when the Alpha/Beta scheme was used, a very large majority of the pupils used number apparatus very infrequently or never. Indeed, even when more modern schemes were in use, most 11-year-olds used number apparatus very infrequently. Thus, the style of mathematics teaching to which many of these upper primary pupils are subjected seems likely to be formal and theoretical.

Further evidence of the very common use of this formal and theoretical style of mathematics teaching is found in a recent study of number and language work in top infant classes by Bennett, Desforges, Cockburn and Wilkinson (1984). They acted as observers in the classrooms of sixteen teachers nominated as better than average by LEA Advisers, and found massive evidence that the major

Conclusion: INSET as an agency of change

ingredient of the top infant mathematics curriculum was a formal coverage of the four rules.

The approach to mathematics teaching found by these two studies is not consonant with the Cockcroft Report's recommendations concerning the primary mathematics curriculum. These include:

> Practical work is essential throughout the primary years if the mathematics curriculum is to be developed in the way which we have advocated ... For most children, practical work provides the most effective means by which understanding of mathematics can develop.
>
> (DES 1982, §289)

> Understanding of place value needs to be developed not only by means of structural apparatus and the abacus but also by using as examples the structure of hundreds, tens and units which underlies both measurement (metres, decimetres and centimetres) and money (pounds, tenpences and pence). ... Many children need further practical experience with structural apparatus so that they can work out for themselves the meaning of large numbers ...
>
> (DES 1982, §299)

Further technological development is taking place. The calculator has become an almost universal tool outside school, and, increasingly, home computers are commonly found in the homes of children of primary school age. There is little indication that the mathematics curriculum in the majority of primary schools has been at all affected by these developments. Trevor Fletcher, HMI, in a discussion paper, *Microcomputers and Mathematics in Schools*, sadly prognosticated that:

> It may be a long time before we can rely on having a sufficient number of primary school teachers with the knowledge and the confidence to implement any general policy of developing programming ability and mathematical knowledge side by side in a systematic way.
>
> (Fletcher 1983)

The present need is to help many primary schools to move forward twenty-five years in their mathematics teaching, as a matter of urgency. However, the only way in which both schools and individual primary teachers can be reached and helped to develop their mathematics teaching is through INSET. It is at least somewhat reassuring that, on occasions when teachers reported major changes in their teaching styles to the researcher after attending INSET courses, these changes were in directions recommended by the Cockcroft Report:

> I am thinking differently, and I have a broader outlook on maths. I am using more apparatus ... I realise how much children need to talk about maths ... and so I encourage children to come and talk about their discoveries ... (§6.3.7)

References

Armstrong, M. (1980). *Closely Observed Children*. London, Writers and Readers.

Assessment of Performance Unit (1980). *Mathematical Development*. Primary Survey Report No. 1. London, HMSO.

Bennett, N., Desforges, C., Cockburn, A. and Wilkinson, B. (1984). *The Quality of Pupil Learning Experiences*. London, Lawrence Erlbaum.

Biggs, E. (1983). *Confident Mathematics Teaching 5 to 13*, Windsor, NFER — Nelson.

Bolam, R. (1979). Evaluating inservice education and training: a national perspective. *British Journal of Teacher Education*. 5(1), 1—15.

Burton, L. (1981). *Strategies and Procedures of Mathematical Problem-solving*. London, Polytechnic of the South Bank.

Department of Education and Science (1978). *Primary Education in England: a Survey by HM Inspectors of Schools*. London, HMSO.

Department of Education and Science (1979). *Matters for Discussion: Mathematics 5—11: a handbook of suggestions*. HMI Series. London, HMSO.

Department of Education and Science (1982). *Mathematics Counts*. Report of the Committee of Inquiry into the teaching of mathematics in schools under the chairmanship of Dr W H Cockcroft. London, HMSO (The Cockcroft Report).

Department of Education and Science (1983). *The In-service Teacher Training Grants Scheme*. DES Circular 3/83. London, DES.

Department of Education and Science and Welsh Office (1978). *Making INSET Work*, London, DES.

References

Donoughue, C. (1981). *In-service: The Teacher and the School*. London, Kogan Page.

Fletcher, T. J. (1983). *Microcomputers and Mathematics in Schools*. London, DES.

Foxman, D. (1984) Personal communication.

Galton, M., Simon, B. and Croll, P. (1980). *Inside the Primary Classroom*. Routledge and Kegan Paul.

Hamilton, D., Jenkins, D., King, C., MacDonald, B. and Partlett, M (1977). *Beyond the Numbers Game*. Basingstoke, Macmillan Education.

Henderson, E. S. (1978). *The Evaluation of In-service Teacher Training*. London, Croom Helm.

McCabe, C. (ed.) (1980). *Evaluating In-Service Training for Teachers*. Windsor, NFER.

Melrose, J. (1982). *The Mathematical Association Diploma in Mathematical Education*. Durham, University of Durham School of Education.

The Open University (1980). PME 233 *Mathematics Across The Curriculum*. Milton Keynes, The Open University.

The Open University (1982). EM 235 *Developing Mathematical Thinking*. Milton Keynes, The Open University.

Osborne, A. (ed.) (1977). *An In-Service Handbook for Mathematics Education*. Reston, Virginia, National Council of Teachers of Mathematics.

Rudduck, J. (1981). *Making the Most of the Short In-service Course*. London, Methuen Educational for the Schools Council.

School Mathematics Project (undated). *In-service kit on fractions*. London, SMP.

Taylor, W. (1978). *Research and Reform in Teacher Education*. Windsor, NFER.

Watts, H. (1981). Can you feed a frog on tadpole food? *Insight: Journal of the National Conference of Teachers' Centre Leaders*. 4,2.

Weindling, D., Reid, M. I. and Davis, P. (1983). *Teachers' Centres: a focus for in-service education*. London, Methuen Educational for the Schools Council.

Index

advertisements, course 76, 80, 97, 181–2
Advisers, LEA, Mathematics *see* Local Education Authorities, Mathematics Advisers
Advisory Council on the Supply and Training of Teachers (ACSTT) 3
Advisory Teachers 138, 169, 186
   in classroom 139–40
Alpha/Beta scheme 190
apparatus 54, 147, 190, 191
applicants for courses, selecting 182
Assessment of Performance Unit surveys 30, 190

Bennett, N., *et al.*, 190–1
Biggs, Edith 6, 179, 190
Bolam, R. 3
Bournemouth conference, *1978* 2–3
British Petroleum, projects 1, 2, 6–7
Burton, L. 124–5

change
   in education system 187–8, 190–1
   in teachers *see* teachers, professional development of
teaching, developments in 47–52, 130–7

classrooms
   advisory teachers in 139–40
   visits to 76–83
Cockcroft Committee 2
   Report 7, 16, 189, 191
     re co-ordinators 170–1, 178
     discussed on courses 21, 22
colleges of education 14
   INSET provision by 1–2, 14–15
   intermittent course for co-ordinators 86–99
   types of 15, 16, 22–4
   preferences 24–7
   and school-based INSET 141–3
communication 128–31, 158–9
computers 17, 103, 191
consolidation stage of development
   for co-ordinators 172–3
   for teachers 154–7
co-ordinators, mathematics 7, 141, 169–70
   courses for 170
     intermittent, at college 86–99
     content 89–91
     mathematics element 92–3
     members 88–9
     monitoring 88
     organisation 86–8
     structure 95–7
   LEA, residential 63–70
   LEA, three days 29–40
   and feedback after courses 174–6

Index

and Head 178—9
needs of 146—8
priorities of, after courses 67—8
professional development of 171—4
role of 65—6, 170—1
courses, INSET
as agency of change 187—8
effectiveness 10—11
effects of *see* effects of courses
feedback on 174—6
needs for, identifying 180
providers of 27—8 180—6
   advertisements 76, 80, 97, 181—2
   planning 181—2
   presentation 185—6
   skills 76
   timing and structure 182—3
reasons for attending 44—5, 57, 77—8, 82—3, 101, 159
in schools 169
and teachers' professional development
   stages 145—6, 163—6
types, colleges' preferences for 25, 26—7
see also monitoring; tutors, course
curriculum 155—6, 157, 158
   Cockcroft re 191
   integration 12—13

Department of Education and Science 5
publications 3, 7, 30, 65—6
Developing Mathematical Thinking (Open University course) 113
discussion 57—8, 183—4
   after classroom visits 79—80, 81, 82—3
   leading 80, 83, 183
   in problem solving 128—9
Donoughue, C. 3—4

effectiveness of INSET 10—11
effects of courses 143—5, 165—6
as agency of change 187—8
co-ordinators 37—9
   feedback after 174—6

follow—up 53, 105—9
infant teachers 60—2
re-entry into school 105—11, 174—7
second teacher from 108—9
environmental work 47—8, 50, 159
equipment *see* apparatus
evaluation 3
exhibitions 104, 175

feedback after courses 174—6
Fletcher, Trevor 191
follow-up after courses
   meeting 53
   in school 105—9
games, mathematical 50, 53, 78
group dynamics 93—4
   leadership 114—15, 120

Hamilton, D., *et al.* 8, 11
handouts 61, 62—3, 65, 70, 95, 175
   lists from 64, 66
   ways to use 184—5
Head teachers 29—30, 142
   and co-ordinators 147—8
   influence of 177—9
   in role-playing exercise 32—3
   and teachers 140—1
Henderson, E. S. 3, 4, 6—7
Homerton College, Cambridge 1

infant teachers, courses for
   3-day residential 55—63
   weekend 43—55
initiation stage
   of co-ordinators 172
   of teachers 153—4
In-service education for teachers (INSET)
   as agency of change 187—8, 190—1
   effectiveness 10—11
   monitoring 1—13, 88, 100
      experiment in 29—40
      methodology 7—10, 28—30
      previous work on 2—6
      proposal 2
      scope of project 6—7
   provision 14—28, 180—6
      level of 189—90
      sources 14—16

Index

types
  colleges 22—7
  LEA Advisers 27—8, 29—40
  preferences among 24—7
  Teachers' Centres 17—33
  see also courses, INSET; schools, INSET work within
integration stage
  of co-ordinators 173
  of teachers 157—9
investigations 65
involvement in courses 58—60

language 128—30, 158—9
leadership
  of discussion 80, 83, 183
  style of 114—15, 120
Local Education Authorities, INSET provision 14—15, 16, 17
  co-ordinators' courses
    residential 63—70
    3-days 29—40
  infant teachers' courses
    3-day 55—63
    weekend 43—55

Mathematics Advisers 14, 38, 39, 63, 138, 186
  courses provided by 27—8, 30, 145—6
  residential centres 42—3
McCabe, C., 3
Making INSET work (DES) 3
Mathematical Association 2
  Diploma in Mathematical Education, courses 2, 90, 149—50
  colleges' preferences for 25
  monitored 5, 9
  statistics re 23—4
mathematics
  content of courses 89—90, 92—3, 96—7
  process of 160
  understanding 183—4
Mathematics across the Curriculum (Open University course) 112—16, 123—4
  INSET course arising from 120—3
  problem solving package from 116—20
Mathematics Advisers, LEA *see* Local Education Authorities, Mathematics Advisers
mathematics co-ordinators *see* co-ordinators, mathematics
Mathematics 5—11 (DES) 30, 64
Melrose, J. 5
Merritt, John 3—4
microcomputers 17, 103, 191
monitoring 1—13, 88, 100
  experiment in 29—40
  methodology 7—10, 29—30
  previous work on 2—6
  proposal for 2
  scope of project 6—7

National Council of Teachers of Mathematics 4—5

Open University 112—24
Mathematics across the Curriculum course 112—16, 123—4
  INSET course arising from 120—3
  problem-solving package from 116—20

personal relationships 122—3
Pinner, Sister Mary Timothy 2, 10
Primary Education in England (DES) 7, 30, 65—6
primary schools
  and secondary, liaison 37, 38
  structure 9
  see also schools
problem solving 112—37, 162
  Open University course and related INSET 113—24
  Mathematics across the Curriculum 114—20
  INSET course from 120—3
  structure and monitoring 113

SPMPS project 124—37
  developments in teaching 130—4
  dissemination 134—5
  language and communication 128—30
  teachers' initial experiences 126—8

recording, by children 129—30, 136
reflection stage of development
  of co-ordinators 173—4
  of teachers 12—13, 160—3

Index

relationships, personal 122—3
reporting back 52—3, 61, 63, 175
residential courses, provision 17, 42—3
role-playing exercise 32—5
Ruddock, J. 4, 95

schemes of work, mathematics, 64, 147—8
  talk from publisher 83
  use surveyed 190—1
schools
  INSET work within, 7, 138—43, 146—51, 169—79, 188—9
  advantages of 25—6
  in classroom 139—41
  college-provided 141—3
  co-ordinators 146—8, 170—4
  courses for individual schools 169
  and teachers' development 148—50
  primary, structure of 9
  primary and secondary, liaison 37, 38
  re-entry into after courses 105—11, 174—7
    for second teacher from same school 108—9
  reporting back to 52—3, 61, 63, 175
  visits to classrooms 76—83
Schools Council 4
Shuard, Hilary 1—2, 10
skills in course-giving 76
Strategies and Procedures of Mathematical Problem-solving (SPMPS) project 124—37
  dissemination 134—5
  language and communication 128—31
  teachers' initial experiences 126—8
  teaching developments 130—4
symbolism, mathematical 130

talks *see* discussion
Taylor, W. 14

teachers
  behaviour as classes 93—4
  in classroom 139—41
  as INSET course providers 186
  needs of, identification of 180
  professional development of 11—13, 148—51, 152—63, 166—8, 187
    consolidation 154—7
    initiation 153—4
    integration 157—9
    match with INSET courses 145—6, 163—6, 180—1
    reflection 160—3
  role of 136—7
  see also Advisory Teachers; co-ordinators; effects of courses; Head teachers; infant teachers; schools, re-entry into after courses; teaching
Teachers' Centres 4, 166, 181
  INSET work of 14, 16, 17
  courses, classroom-based 76—83
  courses mounted by, short 71—6
  single-session at 83—5
  types of provision 17—22, 27
teaching
  developments in 47—52, 130—7
  style of 153, 156
  for adults 76, 94—5
  innovative 112, 123—4
  see also teachers
topic work 48, 50—1, 55—7
tutors, course 75—6, 87—8, 89—91, 93—5, 98, 101, 102—3

United States of America; National Council of Teachers of Mathematics 4—5
visits to classrooms 76—83

Watts, H. 13
Weindling, D., *et al.* 4, 16, 17, 71
written work
  children's recording 129—30
  INSET, teachers' 119—20
  examinations 116